Celestial Sampler

60 Small-Scope
Tours for
Starlit Nights

To my family for their support and encouragement
And to Dudley Observatory for the use of its fine historical library

Celestial Sampler

60 Small-Scope Tours for Starlit Nights

By Sue French

Digger,
Wishing you many clear &
star-filled nights.
Sue French

Foreword by Richard Tresch Fienberg

Sky Publishing Corp.
Cambridge, Massachusetts 2005

© 2005 Sky Publishing Corporation
Published by Sky Publishing Corporation
49 Bay State Road
Cambridge, MA 02138-1200, USA
SkyandTelescope.com

Library of Congress Cataloging-in-Publication Data

French, Sue (Sue C.)
    Celestial sampler : 60 small-scope tours for starlit nights / by Sue French ; foreword by Richard Tresch Fienberg.
        p. cm. – (Stargazing series)
    Includes bibliographical references and index.
    ISBN 1-931559-28-7 (alk. paper)
    1. Astronomy–Observers' manuals. 2. Telescopes–Observers' manuals. I. Title. II. Series.

QB64.F69 2005
522–dc22

2005054721

PRINTED IN CANADA

Contents

Imagine yourself on your own in Paris. It's your first visit to the City of Light, and you want to see as many highlights as you can during your short stay. Standing on a busy street corner, you spy the Eiffel Tower in the distance — and check it off your list. But which Métro stop is closest to the Louvre? Once inside the museum, what's the quickest route to the *Mona Lisa*? Is the Cathedral of Notre-Dame open this afternoon? And where's the Latin Quarter, anyway?

To find the answers to all these questions and more, you buy a Paris guidebook, one with lots of maps. At last you confidently set off to explore the city. Hours later, reflecting on your busy day, you realize that the most fun you had was when, following the instructions in your tourist guide, you strayed from the beaten path, ducked into a charming little café, and had lunch with the locals.

And so it is with stargazing. When you step outside at night and look up, it's easy to find yourself bewildered. Even an experienced backyard astronomer, one used to observing from a light-polluted city or suburb, can feel lost under a star-splashed country sky. There's so much to see — where do you begin? How do you make sure you don't overlook tonight's best celestial showpieces? And where's the Crab Nebula, anyway?

To find the answers, you use a guidebook, one with lots of maps — such as the one in your hands now. It is designed to serve stargazers of all levels, from beginners just learning how to use a telescope to veterans looking for new challenges to test their skills. Its 60 guided tours feature some of the night sky's finest sights, with clear, helpful instructions on how to track them down. It also takes you on numerous side trips to little-known splendors that many stargazers never think to look at.

Celestial Sampler collects 60 installments of a monthly column that has been running in *Sky & Telescope* since July 1999. It started out as Small-Scope Sampler. Former editor in chief Leif J. Robinson got the idea for it when telescope manufacturers introduced a new generation of high-quality instruments with apertures of less than 4 inches (100 millimeters) and stargazers began buying them in droves. The plan was to highlight deep-sky objects — interesting sights beyond the solar system, including double stars, star clusters, nebulae, and galaxies — visible in these scopes under moderately dark skies.

Casting about for an author, Robinson quickly zeroed in on Sue French, a gifted observer from upstate New York whose writing was gaining an avid following among amateur astronomers and whose presentations at star parties were attracting overflow crowds. *Sky & Telescope*'s readers responded just as enthusiastically; Sue's monthly column has become the most popular one in the magazine — and not just among those with small telescopes. Accordingly, it recently underwent a title change to Deep-Sky Wonders. Sue has deservedly taken up the mantle of legendary and beloved observer Walter Scott Houston, who penned a column by that name for nearly 50 years until his death in 1993 at age 81.

Now, unless you live in a rural area, you might be thinking that you haven't much hope of seeing deep-sky wonders from your backyard or driveway because of the pervasive skyglow from artificial lighting. But binoculars and telescopes effectively cut through light pollution, at least to a point. So even if your naked-eye view encompasses only a smattering of the brightest stars, you can still enjoy many hundreds of fascinating deep-sky sights with the aid of optics. And Sue's step-by-step instructions for star-hopping from bright stars to "faint fuzzies" will make doing so both easy and fun.

If you own a large telescope, one whose aperture exceeds 8 or 10 inches, is *Celestial Sampler* not for you? Quite the contrary! With few exceptions, anything that looks good in a small telescope looks better in a big one. I do most of my observing with a 12-inch reflector far from city lights, under beautifully dark skies. I own many books with lists of worthy targets to seek out, and sometimes I'll spend an evening working my way down such a list. But my favorite observing sessions are the ones in which I follow Sue's star-hops, with their eclectic mix of objects from the well known to the obscure.

If you own a computerized Go To telescope, one that catches its celestial prey at the touch of a few buttons, is this book not for you? Again, quite the contrary. Many of the objects featured in *Celestial Sampler* have Messier or NGC designations. Once you've set up your scope and initialized its "brain" according to the manufacturer's instructions, you can call up these sights using the M or NGC buttons on your hand controller. For everything else, you can just enter the coordinates listed in the book's tables. And if your batteries run out, you can easily fall back on good old-fashioned star-hopping!

At the risk of disappointing you, I'm going to let you in on a little secret: In the eyepiece of almost any backyard telescope, most celestial sights other than the Moon and planets look singularly unimpressive. As Sue explains in her introduction, many nebulae and galaxies can't be seen at all unless you give your eyes plenty of time to adapt to the dark and use special observing techniques that require practice to master. And even then they don't look anything like they do in the long-exposure photographs that decorate many telescope boxes and grace the pages of *Sky & Telescope*, other magazines, and most popular astronomy books, including this one.

But here's another tip: Celestial objects are best appreciated not with the eye, but with the mind. Look at the picture of the Crab Nebula on page 44. Beautiful, isn't it? But what's impressive about the Crab is not its color or its intricate lacework of gas filaments — neither of which is visible in most amateur telescopes. What's impressive is that when you look into the eyepiece and discern the Crab's faint, irregular glow, your retina is being tickled by photons that have traveled through space for 6,000 years at 186,000 miles per second. And these photons were energized by a star as massive as the Sun yet only a few miles across and spinning 30 times a second. And this remarkable star, and the nebula enveloping it, are remnants of a cataclysmic stellar explosion witnessed by our ancestors nearly 1,000 years ago.

When you know all that, even the most minute speck of galaxy light in your eyepiece is enough to make your spirit soar and keep you hooked on astronomy forever. But if you don't know all that, the Crab and most other deep-sky objects are just boring fuzzballs that most people wouldn't consider worth a second look.

One of the reasons I enjoy Sue's writing so much is that she doesn't just tell you what to look *at* and what it looks *like* — she also tells you what it *is*. She is no mere astronomical tourist; she is a true explorer, seeking to understand what she observes. Her grasp of the science, and her ability to explain it in simple terms, make her a standout among stargazing writers. With Sue as your guide, you'll be a knowledgeable deep-sky explorer too.

It's a big universe, and this is a small book. I'm sure its 60 sky tours will whet your appetite for more. You have *two* places you can go to get it: Sue's monthly Deep-Sky Wonders column in *Sky & Telescope* and her Take a Star-Hop column in *Night Sky*, our bimonthly magazine for newcomers to amateur astronomy.

And whereas you may not get back to Paris very often, you can return to the night sky — itself a City of Lights — night after night, just by stepping outside with your binoculars or telescope and this book.

Clear skies!

RICHARD TRESCH FIENBERG
August 2005

Welcome to the Night Sky

So you have a small telescope, and you'd like to know what you can see with it? Enough to keep you busy for a lifetime. *Celestial Sampler* will get you started on an amazing cosmic journey and introduce you to a wealth of deep-sky wonders that will entertain you for years to come. Over time, these treasures will become old friends who will reveal new secrets as you gain observing experience.

Most of the objects presented in this book are visible in a 4-inch (100-millimeter) telescope under a moderately dark sky — a sky dark enough for you to spot the seven stars of the Little Dipper with your unaided eye. But many can also be seen in light-polluted skies. These celestial sights fall into several different classes of objects.

Variable Stars. Many stars undergo changes in brightness that can be followed with a telescope. Most of the variable stars that I mention in this book are *pulsating variables,* stars whose surface layers periodically expand and contract. I've also included a few examples of *eclipsing binaries,* pairs of stars whose orbital plane lies nearly along our line of sight. This means we see the stars periodically eclipse (hide) one another, causing a decrease in the apparent brightness of the binary. The most famous eclipsing binary is Algol — Beta (β) Persei — in the constellation Perseus.

Multiple Stars. Two or more stars often appear very close together in the sky. Some are merely *optical doubles* — stars whose components look close but actually lie at different distances and are not physically related. However, most multiple-star systems are bound by gravity and orbit each other in dances that may be quite complex. When the components of a pair are separated by enough space that we can split them with a telescope, we call it a *visual double.* The brighter star of a pair is called the *primary* or A star, while the companion is usually called the *secondary* or B star. If there are more than two stars, they are designated by succeeding letters in order of diminishing brightness.

A beautiful example of a double star is Albireo in Cygnus, the Swan.

Star Clusters. Larger groups of stars are known as star clusters. *Open clusters* are groups of stars that were born together in space and are loosely bound by gravity. They gradually lose their stars over millions of years. *Globular clusters* are long-lived groups held in a tight gravitational embrace. While an open cluster may have tens to thousands of stars, globular clusters typically hold more than 100,000.

Nebulae. Clouds of gas and dust that inhabit regions between and around stars are called nebulae. There are several different types. *Emission nebulae* are heated by the radiation of nearby hot stars and glow with their own light. Most emission nebulae are stellar birthplaces, but two special types are associated with the death of a star. *Planetary nebulae* are composed of material being puffed off by an aged star as it is running out of nuclear fuel. However, massive stars don't die as gracefully. Many end their lives in titanic explosions known as *supernovae.* They violently shed material into space that can sometimes be observed as a glowing *supernova remnant.*

Some nebulae do not emit their own light. *Reflection nebulae* are seen when particles of dust in the nebula reflect the light from nearby stars. If there are no stars in the nebula's vicinity, it remains dark. Dark nebulae are only seen in silhouette if they happen to blot out a rich field of distant stars or are projected against a bright background nebula.

Galaxies. A *galaxy* is an immense collection of gas, dust, and billions of stars. The stars, clusters, and nebulae you can see with a small telescope are, with very few exceptions, members of our own Milky Way Galaxy. But there are many other galaxies far beyond the twinkling stars that adorn our sky. Hundreds lie within the grasp of a small telescope, though most will appear only as softly glowing fuzzballs. The easiest galaxy to find is the Great Andromeda Galaxy; it's a naked-eye sight under a dark sky in the autumn constellation of Andromeda.

The Sagittarius section of the Milky Way is rich in deep-sky wonders. It's a great place for novice stargazers to begin sampling what the heavens have to offer.

Galaxies abound in and around Markarian's Chain, from M84 (right) to NGC 4477 (upper left). This string of galaxies straddles the Virgo–Coma Berenices border. North is toward the upper left in this 2°-wide field. To explore this galaxy chain, see page 84.

Using a Star Chart

To locate the objects in this book, you'll need to navigate the included star charts, perhaps in conjunction with an atlas of your own. You'll find 12 monthly all-sky maps beginning on page 20, while each essay contains a large-scale finder chart.

We use latitude and longitude to specify the position of a place on Earth, and these coordinates let us find that place on a world atlas. In the sky, the equivalent of latitude is called *declination* (abbreviated as *dec.*) and is measured in degrees north and south of the *celestial equator* (an extension of the Earth's equator into the heavens that splits the sky in half). Each degree is divided into 60 *arcminutes* (60'), and each arcminute is divided into 60 *arcseconds* (60"). The declination of the brilliant winter star Sirius would be written –16° 42' 47" or, using decimal arcminutes, –16° 42.8'. The minus sign indicates that Sirius is below the celestial equator.

The equivalent of longitude is called *right ascension* (abbreviated as *RA*) and gives us an east-west measure in the sky. Unlike longitude, however, right ascension is not measured in degrees. As the Earth rotates, the stars seem to make a full east-to-west circuit of the sky in almost exactly 24 hours. It then seems quite natural to divide the sky into 24 hours of right ascension. Each hour is divided into 60 minutes (60^m), and each minute is divided into 60 seconds (60^s). The right ascension of Sirius would be $6^h 45^m 09^s$ or $6^h 45.2^m$.

Don't be intimidated by these numbers. Although I list the right ascension and declination of each object I discuss (so those with computerized telescopes can simply enter the coordinates), I also tell you how to find each celestial sight by *star-hopping*. This is a process that involves beginning at an easy-to-find object or bright star, and then (with the aid of the star maps in each essay) "hopping" from one recognizable star or pattern of stars to another until you reach your target.

Navigating the Sky

The point in the sky toward which Earth's north pole points gives us the sky's north pole. The *north celestial pole* happens to lie very close to Polaris, the North Star, which is the brightest star of the Little Dipper (see page 88). On a star chart, north is always toward this spot in the sky. If you are looking at a star map where north is up, then west is to the right. This is contrary to maps of Earth for a very simple reason — you look down at

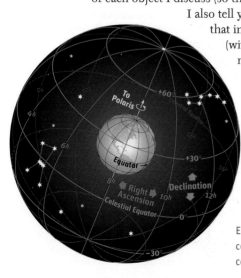

Earth is at the center of the *celestial sphere* — an imaginary sky surface in which celestial objects appear to be embedded. Lines of latitude and longitude on Earth correspond to lines of declination and right ascension on the celestial sphere.

Earth, but up at the sky.

To use the small finder charts that appear in every essay, you need to figure out how the directions in your eyepiece or finder compare to the directions on the map. One simple method is to note which way a star in the center of the field is drifting: that direction will be west. (If your telescope has a motor drive, turn it off.) If you have a telescope that gives an *erect* (normal) or *inverted* (upside-down) image, north will be clockwise from there. If your scope gives a *mirror-image* view, north will be counterclockwise. If you're not sure, give your scope a nudge toward Polaris to see which way north lies.

You must also master the concept of angular distance to navigate the night sky, but that is much simpler than it sounds. Remember right ascension and declination? Since the night sky looks like a huge dome with stars stuck on its inside surface, the right ascension and declination of any object gives its position on this great *celestial sphere*. But the distance between two spots (or the width of any object) on that dome is expressed as an angle — the angle you'd have to swing your eye through to look from one to the other. For example, the Moon is about ¹/₂° across, the Big Dipper is 25° long, and from the horizon to the *zenith* (the point straight overhead) is 90°.

It is fairly simple to figure out the field of view of each of your eyepieces, though the terms — apparent and true field of view — can be confusing. The *apparent field* is the angle you'd have to swing your eye through to look from one edge of the eyepiece's field to the other. The *true field* is the amount of sky (in degrees or arcminutes) that is actually seen when the eyepiece is used in your telescope.

Manufacturers generally give the apparent field of their eyepieces. *The true field of an eyepiece is approximately equal to the apparent field divided by the magnification that eyepiece gives you.* If you don't know the magnification of your eyepiece, divide the focal length of your telescope (a large number usually printed somewhere on the telescope or listed in the scope's manual) by the focal length of the eyepiece (the number on the eyepiece itself).

For example, if you have an eyepiece with an apparent field of 50° that gives you a power of

To estimate how far apart things are in the sky, hold your hand at arm's length in front of you. Your little finger is about 1° wide; your three middle fingers together (in a Boy Scout salute) cover roughly 5°, the width of your fist is about 10°, and your wide-open hand is about 20° from the tip of your thumb to the end of your little finger.

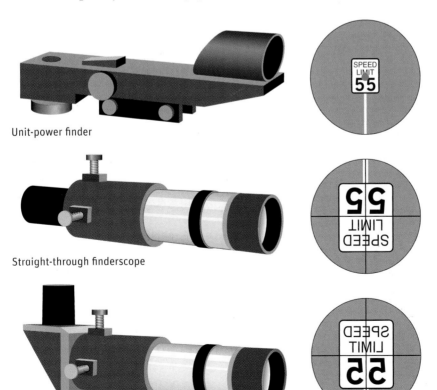

Unit-power finder

Straight-through finderscope

Right-angle finderscope

Most of the large-scale charts that accompany each essay include a circle illustrating a finderscope's field of view. If you're trying to match the chart to what you see in the sky, remember that most straight-through finderscopes *(middle)* show an inverted view — they turn the image upside down. Most right-angle finderscopes *(bottom)* give a mirror-image view — they mirror-reverse the image (left to right). Unit-power finders *(top)* don't magnify the sky at all.

50×, the true field in your scope is 1° (50°/50 = 1°). An eyepiece with an apparent field of 50° that gives you 100× yields a true field of ¹/₂° (50°/100 = ¹/₂°). Want to confirm your calculations? Look at the Moon. Your eyepiece has a true field of about ¹/₂° if the Moon just barely fills it.

You now have all the tools you need to use a star chart — position, direction, and distance. To begin, I generally aim my scope at the nearest star I can see. The star cluster M41 is 4° south of Sirius, so I'd aim my scope at Sirius and sweep 4° south. It sometimes helps to use an eyepiece with a 1° true field so that you can easily count off the number of degrees you are moving. Keeping an eye on distinctive star patterns along the way often helps lead you to your target.

If there are no bright stars nearby, a finder can help you locate your quarry. An *optical finder-scope* shows a larger area of the sky than your main scope, and you can use it to hop from a bright star to a dimmer one near your quarry. *Unit-power* or *reflex finders* (sighting devices that seem to place a bull's-eye or dot on the sky) will let you accurately aim your scope at a point between the visible stars — a spot you can pick by examining the star chart.

Moving in the correct direction is simple if you have an equatorial mount. Aligning your telescope by pointing the mount's polar axis toward the North Star lets your scope move along the cardinal directions. But if you have an altazimuth mount, which moves up and down and back and forth (like the popular Dobsonian), it's easy to lose track of directions. To keep them straight, rotate your star chart until it matches the view in your scope. This will give you a better idea of which way to move.

Finally, the all-sky maps and finder charts in *Celestial Sampler* will take you only so far. Eventually you'll need a proper star atlas. With it you'll discover many more celestial delights than I can discuss in this book. Turn to the Resources section (page 161) for a short list of some of the better atlases.

How Bright, How Far

The brightness of any celestial object is called its *magnitude.* The Greek astronomer Hipparchus created the magnitude scale in the 2nd century BC. He called the brightest stars he could see "first magnitude" and the faintest "sixth magnitude." As a result, and contrary to what you might expect, the higher the magnitude number, the fainter the object. With the invention of the telescope, observers spotted even fainter stars; today astronomers study objects as dim as magnitude 30. At the other end of the scale, it turned out that some of Hipparchus's first-magnitude stars were brighter than others, so the scale was extended the other way into negative numbers. Hence Sirius, the brightest star in the night sky, is magnitude –1.4, Venus is even brighter (usually magnitude –4), and the full Moon shines at magnitude –13.

Another term you'll see, particularly in the tables included in each essay, is *light-year.* This is the distance that light travels in a year: 9.5 trillion kilometers or 5.9 trillion miles. A light-year is a measure of distance, not time. Most of the stars we see are dozens or hundreds of light-years away. Many nebulae and globular clusters are thousands of light-years distant. The Andromeda Galaxy, the nearest large galaxy to our Milky Way, is some 2.5 million light-years away. Other galaxies are many millions, even billions of light-years out. It's a big universe!

Observing Tips

Here are a few tips that will help you make the most of your deep-sky forays.

The Speed of Dark. To observe faint objects, you need to let your eyes become adapted to the dark. One change occurs rather rapidly when you step out under the night sky — the pupil of your eye opens wide and allows more light to enter. But this is a minor improvement compared to what happens next. A slow chemical change within your eyes makes them much more sensitive to light. This takes about 30 minutes, after which you have reasonably good night vision. (Full adaptation takes about two hours.)

Since exposure to light will quickly ruin your dark adaptation, how can you read your charts or

A tube extending beyond the front lens of your refractor or compound telescope helps prevent dew from condensing on your scope's optics.

take notes at the telescope? There are two approaches. The receptors that give your eyes good night vision aren't particularly sensitive to red light, so using a dim red-light flashlight will help preserve your night vision. A second tack is to wear a patch over your observing eye when you're not looking through the eyepiece. You can use the other eye when employing a flashlight to read charts, write notes, or change eyepieces. It's still a good idea to keep the light dim, though.

Dewy, Dewy Night. All manner of dew zappers and dew shields are now commercially available for your telescope. Low-tech methods for preventing dew accumulation include keeping the items in question covered or slightly warmer than the night air. For example, you can keep your eyepieces in a case and close it between uses, or you can keep them (preferably covered with dust caps) in your pockets. But what about your increasingly soggy atlas? The simplest method I've found to protect your star chart is to put a sheet of plexiglass on top of it. This also keeps the pages from blowing around on a windy night.

Observing in the Hood. Stray light shining into your eyes from the surrounding environment can be distracting and can hamper telescopic views. Throwing a dark cloth over your head can shield you from unwanted illumination and make it easier for you to find those really faint fuzzies. Hoods made specifically for this purpose are commercially available. Just be careful not to breathe on your eyepiece while you're under there. (On the other hand, don't hold your breath either. Your eyes and brain need oxygen to function at peak efficiency.)

At the Bottom of the Sea. We observe the heavens from beneath a turbulent ocean of air that often distorts the images we try so hard to capture. When stars seem to shimmer and refuse to stay sharply focused, we say that the *seeing* is poor. Good seeing is needed to pick out fine details in celestial objects. Short of moving to a place that offers steadier skies, you can't do much about atmospheric seeing. However, local seeing conditions are easier to control.

If your telescope is stored indoors, it needs time to reach the outdoor temperature before it can give good images. Otherwise, heat transfer will create air currents in the scope's tube that will soften your images and distort your view. It's a good idea to put your scope outside at least 30 minutes before you start observing. If possible, pick an observing site where you won't have to observe over large heat radiators such as roofs, concrete, or blacktop. Observing in the early morning hours (when these objects have already shed most of their excess heat) will also help.

Transparency is another atmospheric condition that affects your views. Haze, dust, and pollutants can dim the sky and make it difficult to view faint objects. With poor transparency, you need to concentrate on bright objects or wait for a clearer night.

Light pollution is wasted light shed needlessly into the sky, where it serves no useful purpose.

This image sequence shows the double star Zeta Aquarii being affected by thermal distortion in the tube of a reflector. Letting your scope reach the outdoor temperature before using it helps alleviate this problem.

Here's why you should try to observe from as dark a site as possible. The images were taken from a city suburb *(left)* and a rural location. Bright stars show through regardless, but faint stars and the Milky Way are lost in a city sky.

I've included plenty of color images in this book, but don't expect to see a Technicolor universe in your eyepiece. Planets and some stars show limited hues, but colors in faint objects materialize only in time-exposure images like the one at left. To your eye, most non-stellar deep-sky objects will appear only as grayish glows, though the Orion Nebula and a few other bright ones show some color in big scopes. (The right-hand image has been altered to resemble a telescopic eyepiece view.)

Instead, it turns the velvet black of night into a dingy gray. It's tough to trace out the soft radiance of a gossamer nebula or an ethereal galaxy against a gray sky. This is a circumstance where small scopes are at an advantage. They are easier to pack up and take to a better observing site. Even at a dark site, the annoying glare of a solitary light nearby can hinder the view. Observing hoods can be put to good use here.

Power Corrupts . . . but only if you use too much. Inexpensive scopes are sometimes sold with ludicrous power claims. Any scope can be supplied with a super-high-power eyepiece, but overdoing it just gives you a view that is too faint and blurry to be useful. On an average night in the semi-rural area where I live, the maximum useful magnification is about 30× per inch of aperture. So my 4.1-inch refractor should handle 123× and my 10-inch reflector should be fine at 300×. On a steady night, 50× to 60× per inch of aperture is easily attainable with the 4.1-inch, but the 10-inch still hits its limit at 300×. The problem is that the seeing here is seldom good enough to support more than 300× no matter what telescope I use.

The easiest way to find the optimum power for viewing a particular object on any given night is to sneak up on it. Start with your lowest-power eyepiece, then step up to your next higher, and so on. When you think the image is starting to look too faint or blurry, drop back to the last eyepiece you used and study the object carefully. Novices tend to favor lower magnifications than more experienced observers do.

If an object is very small and dim, you may not be able to see it at your lowest power. When observing a non-stellar object, observers often perceive an increase in contrast with increased magnification. This is an illusion. Spreading its light over a larger area has dimmed the background sky, but the nebula or galaxy has been dimmed by the same amount for the same reason. What has really happened is that you've made the task easier for your eyes because they're much better at detecting large faint objects than small faint ones.

Increasing contrast by increasing magnification does work on point sources of light. If you magnify a star cluster, you will be decreasing the apparent brightness of the background sky. How-

> 30 Nov 2004   10:04 p.m.
> Comet Machholz   C/2004 Q2
> 80 mm refractor @
>   60X
> about 6th magnitude
> diffuse bloppy head
> short tail, but Moonlight washing
> most detail out. 8x56 binocular
>   view is a
>   fozzy patch of
>   light in an
>   empty part of
>   sky just a
>   field below &
>   west of epsilon
>   Lepus.
>
> Swan band filter not much
> improvement
> Next week when the Moon is not a
> factor should make the comet a
> naked eye object.
> Temp outside 15° with 2"
> fine snow powder on ground.
> Millenium Star Atlas page 352
> near ngc 1744

Keeping a sky diary is a good way to remember what you've seen. It doesn't have to be fancy; a regular notebook will do.

ever, stars are much too small to have their size increased by magnification. They remain point sources at any power and thus stand out better against a darker background. Just don't go overboard; due to seeing and the nature of telescope optics, stars will start to bloat at very high powers and there will be little benefit from a further increase in magnification.

Tricks of the Trade

A number of techniques can help you make the most of your views through a telescope.

The wiggle factor takes advantage of the fact that it's easier to see a faint moving object than a faint stationary one. If you're pretty sure your quarry is in the field of view but you still can't see it, try gently tapping the side of the telescope's tube. The imparted wiggle just might make the difference.

Averted vision is the practice of looking off to the side of a faint object instead of directly at it. This lets the light fall on a more sensitive region of your eye. Averted vision can help bring out detail in deep-sky objects or even make it possible to detect an object that is otherwise invisible.

Slightly defocusing a star can make its color easier to detect. This comes in handy when trying to compare the components of multiple stars.

Finally, don't underestimate the importance of experience. Observe, and observe often. The more your eye and brain learn to work together under the somewhat unusual conditions at the eyepiece, the more you will see.

My Personal Approach

Sadly, it's not always clear when I want it to be. Unexpected clouds often belie predictions of clear nights. Lengthy spells of inclement weather have left me with lists of unobserved objects whose season has passed. Years of this have led me to a simple plan.

I keep a notebook that's divided into sections by hours of right ascension. Whenever my reading turns up an interesting object, I enter it into the appropriate section with its name, type, position, constellation, and a chart number (from a suitable atlas). I often note which scope I'd like to observe it with. Now when a clear night appears, all I need do is grab my notebook. Opening it to whatever section is well placed in the sky, I have a ready-made observing plan.

This book already has its deep-sky wonders arranged by month, so the work has been done for you. But I highly recommend this method when you expand your observing to treasures not included here.

Taking notes at the eyepiece has made me a much better observer. I write my observations in a small notebook, but some folks prefer dictating to an audio recorder. My notes include a description of the object; telescope, eyepieces, and filters used; date and time; and sky conditions. Later I transfer these observations to 5-inch by 8-inch index cards filed by right ascension. Those with a more high-tech bent often keep their observational logs on their computers.

Writing a description or penciling a sketch forces me to pay attention and look for details. Having my logs to look back on lets me see how my abilities have changed, as well as what objects and features still await discovery.

Another way to get more from observing is to share the view. My astronomy club holds frequent public star parties. Watching someone else get excited about an object that I've viewed a hundred times breathes new life into it. So even if you enjoy the peaceful calm of a night alone under the stars, occasionally take a little time to show the wonders of the universe to someone else. You'll be paid back manyfold.

Finally, let me leave you with these encouraging words from Leland S. Copeland, *Sky & Telescope's* first Deep-Sky Wonders columnist in the 1940s.

The earth's great shadow sweeps around
And ends the fulgent day;
The mountains doff their twilight blue,
The lowlands lose their gray.
But overhead the stars return,
The cloven Milky Way,
And eyes that look through telescopes
Amazing things survey.

SUE FRENCH
Scotia, New York

How to Use the All-Sky Star Maps

Facing NW

Page 28

Facing NW

Even if you're a novice stargazer, you can usually find the Big Dipper or Orion without too much trouble. But what about some lesser-known star patterns such as Delphinus, Sagitta, or Monoceros?

If you're not familiar with the night sky, the 12 all-sky star charts (one for each month) beginning on page 20 will help you locate the bright stars and major constellations visible this evening or any time during the year when it's dark. The charts may look daunting, but they're nothing more than a representation of the starry dome, flattened and shrunk onto a page.

Getting Started

Leaf through the next few pages of this book and find the all-sky chart that's good for the date and time you want to observe. Make sure you use the star map within an hour of so of the listed times. (Note that standard time is used throughout. If you're on daylight-saving time, don't forget to add one hour to the listed times.)

You may have to flip back and forth between months if your observing time is late in the evening (or after midnight). For example, to see how the sky appears in early August at 1 a.m. daylight-saving time, (midnight standard time), you'll want to use the October star chart.

When you venture outside to stargaze, you'll need to know in which direction you're looking. (If you aren't certain, note where the Sun sets — that's west; north is to your right.) Hold the map out in front of you and turn it so the label along the curved edge that matches the direction you're facing is right-side up. That curved edge represents the horizon, and the stars above it are now oriented to match the sky. The map's center is the *zenith*, the point in the sky directly overhead.

At the lower left on each chart is a scale depicting the *magnitude* (brightness) of the stars. The 12 all-sky maps show stars to magnitude 4.5, about as faint as you can see from a suburban environment. (See page 14 for a brief discussion of magnitudes.) At lower right is a key to the different types of stars and deep-sky objects plotted on the charts. These sights and many others are described in the chapters that follow, where the inclusion of more detailed finder charts with each essay makes it easy to locate the various celestial wonders.

Finding the Big Dipper

As an example of how to use these all-sky charts, turn to page 28 and locate the Big Dipper. Rotate the page and hold it so the "Facing NW" label is right-side up. The Big Dipper will be about one-quarter of the way between the horizon and the zenith. (It looks like a giant spoon with three stars in its Handle emerging from the left of a bowl-shaped group of four stars.)

Now go outside around one of the dates and times listed on page 28, face northwest, and look one-quarter of the way up from the horizon. Assuming that no trees, houses, or clouds block your view, you'll see the seven stars of the Big Dipper.

Tips for Success

When you first step outside, look for only the brightest stars on the map — those shown with the biggest dots. Initially, at least, ignore the fainter ones, particularly if you live in a city or suburb (or have a bright Moon in the sky), because they'll be invisible through all the light pollution. Also remember that there's a much bigger difference between the bright and faint stars in the real sky than is suggested by the charts.

Something else to keep in mind is that the star patterns will look much larger in the sky than they do on paper. Try this experiment to see how much larger. Find the Big Dipper in the April all-sky map on page 23, and then hold the chart at arm's length in front of you. Fully extend your other arm with your fingers splayed as wide open as possible. From the tip of your thumb to the tip of your little finger covers about 20° of sky — a little less than the width of the Big Dipper in the sky. Now look at the size of the Dipper on the chart . . . rather tiny by comparison.

These all-sky charts are drawn for skygazers who live anywhere in the world between 35° and 45° north latitude. But even if you don't live within this region, the charts will still be useful. If you're south of 35° N, stars in the southern part of the sky will appear higher than the map shows, and stars in the north will be lower. If you live north of 45° N latitude, the reverse will be true.

Although many of the brighter Messier deep-sky objects are plotted on the charts, you won't find markers for the much-brighter planets. That's because the planets always change their positions. Instead, note the green line cutting through each chart. It's called the *ecliptic* — the path along which the Sun, Moon, and planets travel. If you notice a bright "star" near the ecliptic that's not on the chart, you've spied a planet.

Greek Letters on Star Maps

On both the large all-sky maps and the small star charts found throughout the rest of this book, many of the stars in each constellation are identified with Greek letters. A constellation's most brilliant star is usually called Alpha, the first letter in the Greek alphabet; the second brightest is Beta, and so on.

α	Alpha	ν	Nu
β	Beta	ξ	Xi
γ	Gamma	o	Omicron
δ	Delta	π	Pi
ε	Epsilon	ρ	Rho
ζ	Zeta	σ	Sigma
η	Eta	τ	Tau
θ	Theta	υ	Upsilon
ι	Iota	φ	Phi
κ	Kappa	χ	Chi
λ	Lambda	ψ	Psi
μ	Mu	ω	Omega

Numerous bright stars have Arabic names that have remained in common usage. For instance, while the brightest star in Cygnus is Alpha (α) Cygni, it's better known as Deneb.

Star magnitudes
−1 0 1 2 3 4

Variable star Double star Open cluster Globular cluster Galaxy Planetary nebula Diffuse nebula

Facing South

WHEN TO USE THIS MAP
JANUARY EVENINGS

Late January   Dusk
Early January   8 p.m.

These are standard times.
If daylight-saving time is in effect, add one hour.

OTHER TIMES

Late December	9 p.m.
Early December	10 p.m.
Late November	11 p.m.
Early November	Midnight
Late October	1 a.m.
Early October	2 a.m.
Late September	3 a.m.
Early September	4 a.m.

Star magnitudes
−1 0 1 2 3 4

Variable star · Double star · Open cluster · Globular cluster · Galaxy · Planetary nebula · Diffuse nebula

WHEN TO USE THIS MAP

FEBRUARY EVENINGS

| Late February | Dusk |
| Early February | 8 p.m. |

These are standard times.
If daylight-saving time is in effect, add one hour.

OTHER TIMES

Late January	9 p.m.
Early January	10 p.m.
Late December	11 p.m.
Early December	Midnight
Late November	1 a.m.
Early November	2 a.m.
Late October	3 a.m.
Early October	4 a.m.

Star magnitudes
−1   0   1   2   3   4

Variable star
Double star
Open cluster
Globular cluster
Galaxy
Planetary nebula
Diffuse nebula

WHEN TO USE THIS MAP
MARCH EVENINGS

Late March	Dusk
Early March	9 p.m.

These are standard times.
If daylight-saving time is in effect, add one hour.

OTHER TIMES

Late February	10 p.m.
Early February	11 p.m.
Late January	Midnight
Early January	1 a.m.
Late December	2 a.m.
Early December	3 a.m.
Late November	4 a.m.
Early November	5 a.m.

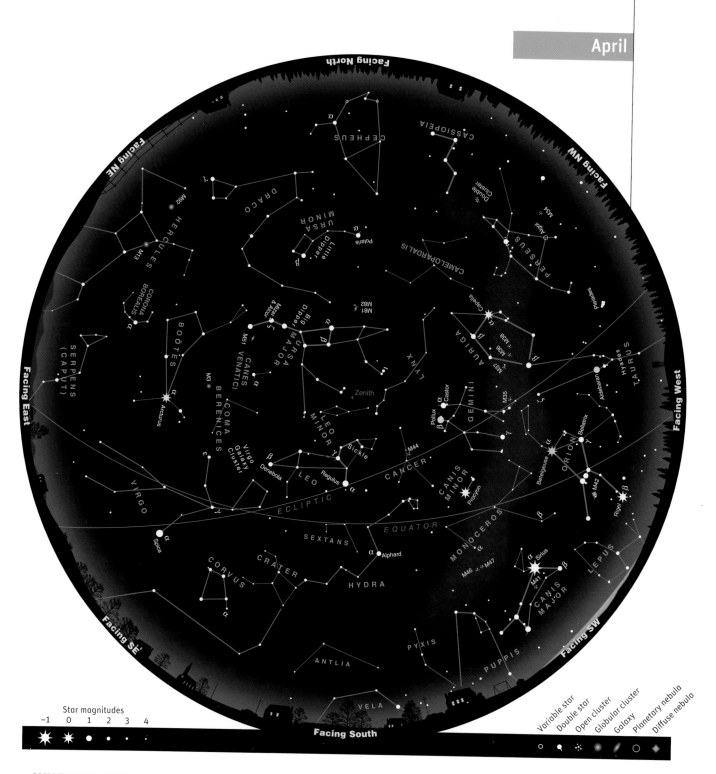

Star magnitudes
−1 0 1 2 3 4

Facing South

Variable star · Double star · Open cluster · Globular cluster · Galaxy · Planetary nebula · Diffuse nebula

WHEN TO USE THIS MAP

APRIL EVENINGS

Late April	Dusk
Early April	9 p.m.

These are standard times.
If daylight-saving time is in effect, add one hour.

OTHER TIMES

Late March	10 p.m.
Early March	11 p.m.
Late February	Midnight
Early February	1 a.m.
Late January	2 a.m.
Early January	3 a.m.
Late December	4 a.m.
Early December	5 a.m.

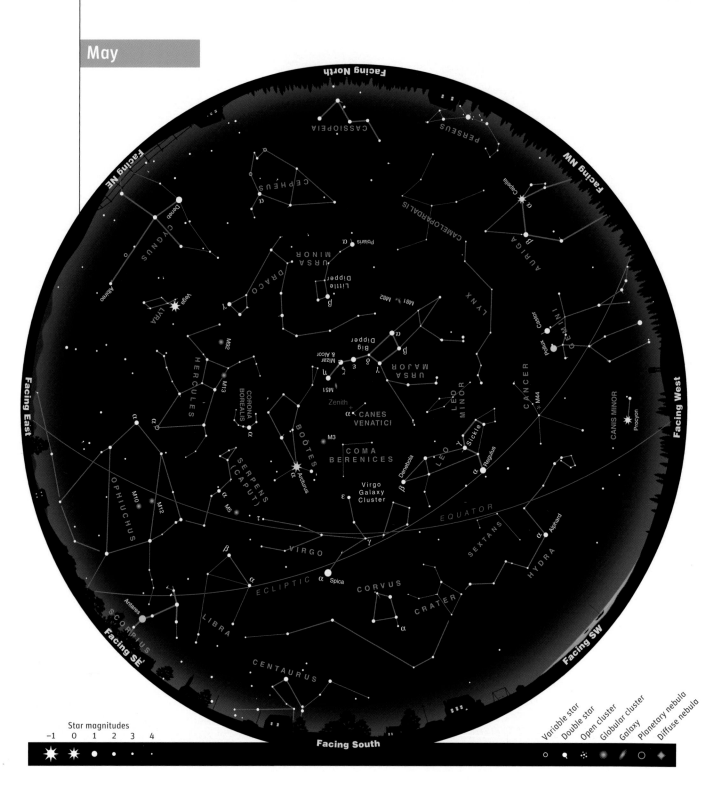

Star magnitudes
-1 0 1 2 3 4

Facing South

Variable star · Double star · Open cluster · Globular cluster · Galaxy · Planetary nebula · Diffuse nebula

WHEN TO USE THIS MAP

MAY EVENINGS

Late May   Dusk

Early May   10 p.m.

These are standard times.
If daylight-saving time is in effect, add one hour.

OTHER TIMES

Late April	11 p.m.
Early April	Midnight
Late March	1 a.m.
Early March	2 a.m.
Late February	3 a.m.
Early February	4 a.m.
Late January	5 a.m.

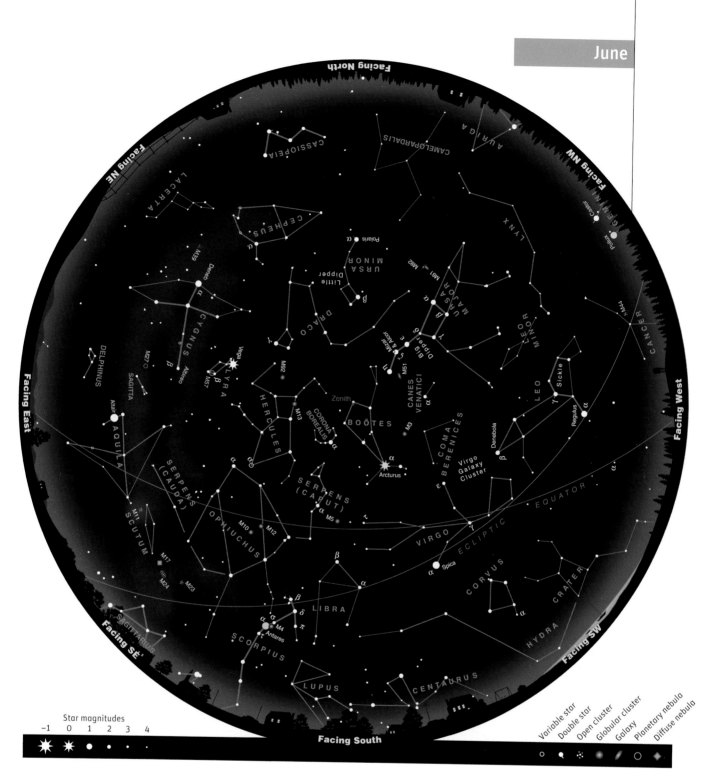

Star magnitudes
−1   0   1   2   3   4

Variable star | Double star | Open cluster | Globular cluster | Galaxy | Planetary nebula | Diffuse nebula

WHEN TO USE THIS MAP

JUNE EVENINGS

| Late June | Dusk |
| Early June | 10 p.m. |

These are standard times.
If daylight-saving time is in effect, add one hour.

OTHER TIMES

Late May	11 p.m.
Early May	Midnight
Late April	1 a.m.
Early April	2 a.m.
Late March	3 a.m.
Early March	4 a.m.
Late February	5 a.m.

Star magnitudes
−1 0 1 2 3 4

Variable star   Double star   Open cluster   Globular cluster   Galaxy   Planetary nebula   Diffuse nebula

Facing South

WHEN TO USE THIS MAP
JULY EVENINGS

Late July	Dusk
Early July	10 p.m.

These are standard times.
If daylight-saving time is in effect, add one hour.

OTHER TIMES

Late June	11 p.m.
Early June	Midnight
Late May	1 a.m.
Early May	2 a.m.
Late April	3 a.m.
Early April	Dawn

WHEN TO USE THIS MAP

AUGUST EVENINGS

Late August	Dusk
Early August	9 p.m.

These are standard times.
If daylight-saving time is in effect, add one hour.

OTHER TIMES

Late July	10 p.m.
Early July	11 p.m.
Late June	Midnight
Early June	1 a.m.
Late May	2 a.m.
Early May	3 a.m.

September

Facing North

Facing NE

Facing NW

Facing East

Facing West

Zenith

Facing SE

Facing SW

Facing South

Star magnitudes
−1 0 1 2 3 4

Variable star Double star Open cluster Globular cluster Galaxy Planetary nebula Diffuse nebula

WHEN TO USE THIS MAP
SEPTEMBER EVENINGS

Late September   Dusk
Early September   8 p.m.

These are standard times.
If daylight-saving time is in effect, add one hour.

OTHER TIMES

Late August	9 p.m.
Early August	10 p.m.
Late July	11 p.m.
Early July	Midnight
Late June	1 a.m.
Early June	2 a.m.
Late May	Dawn

Star magnitudes
−1 0 1 2 3 4

Variable star · Double star · Open cluster · Globular cluster · Galaxy · Planetary nebula · Diffuse nebula

WHEN TO USE THIS MAP
OCTOBER EVENINGS

Late October	Dusk
Early October	8 p.m.

These are standard times.
If daylight-saving time is in effect, add one hour.

OTHER TIMES

Late September	9 p.m.
Early September	10 p.m.
Late August	11 p.m.
Early August	Midnight
Late July	1 a.m.
Early July	2 a.m.
Late June	Dawn

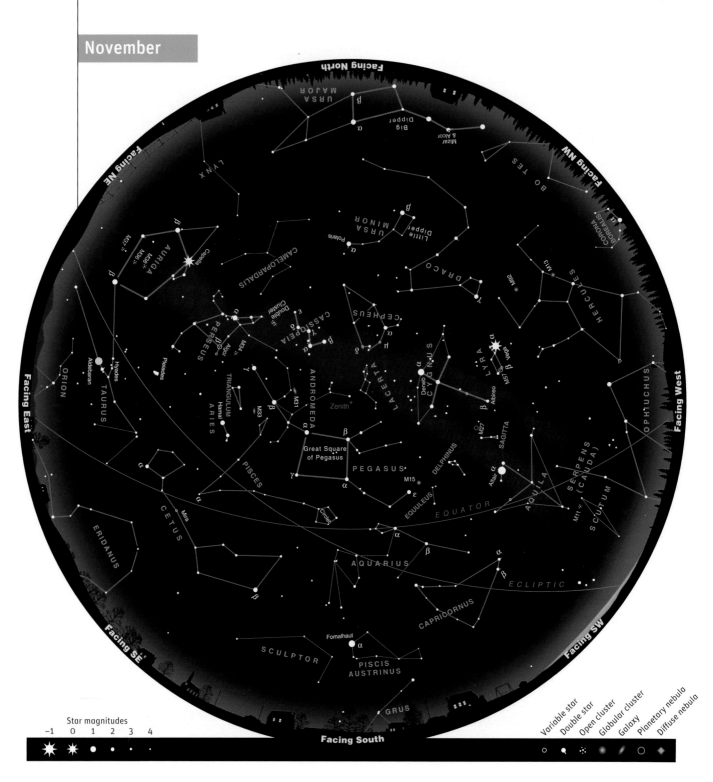

WHEN TO USE THIS MAP

NOVEMBER EVENINGS

Late November	7 p.m.
Early November	8 p.m.

These are standard times.
If daylight-saving time is in effect, add one hour.

OTHER TIMES

Late October	9 p.m.
Early October	10 p.m.
Late September	11 p.m.
Early September	Midnight
Late August	1 a.m.
Early August	2 a.m.
Late July	3 a.m.

Star magnitudes
−1 0 1 2 3 4

Variable star   Double star   Open cluster   Globular cluster   Galaxy   Planetary nebula   Diffuse nebula

30 ★ *Celestial Sampler*

Star magnitudes
−1 0 1 2 3 4

Variable star · Double star · Open cluster · Globular cluster · Galaxy · Planetary nebula · Diffuse nebula

WHEN TO USE THIS MAP
DECEMBER EVENINGS

Late December	7 p.m.
Early December	8 p.m.

These are standard times.
If daylight-saving time is in effect, add one hour.

OTHER TIMES

Late November	9 p.m.
Early November	10 p.m.
Late October	11 p.m.
Early October	Midnight
Late September	1 a.m.
Early September	2 a.m.
Late August	3 a.m.

January • February • March

Chapter

1

The winter sky is ablaze with brilliant stars and deep-sky sights aplenty. Of course you'll want to explore old favorites like the Sword of Orion and the Pleiades in Taurus, but I'll also point you toward some lesser-known celestial wonders. From Auriga and Gemini in the north to Canis Major and Puppis in the south, the winter sky contains numerous open star clusters — delightful sights in small telescopes.

A Hero's Quest

IN GREEK MYTHOLOGY Cassiopeia was a vain queen who claimed she was more beautiful than the daughters of Nereus, the sea god. This boast began a chain of events leading Queen Cassiopeia and her husband, King Cepheus, to sacrifice their daughter Andromeda to the sea monster Cetus. Fortunately for Andromeda, the hero Perseus happened by on Pegasus, the Winged Horse, and slew the monster. All the characters in this story are honored with constellations high in the western and northern sky at this time of year. You can find them on our all-sky map on page 20. Perseus is at the zenith, the map's center.

Perseus is usually portrayed carrying the head of Medusa, which is marked by the 2nd-magnitude star **Algol.** In some versions of the myth, Perseus killed Cetus by showing him Medusa's head — a sight so hideous that anyone (apparently even a sea monster) who saw it would turn to stone. The name Algol comes from the Arabic *Ra's al Ghul,* "the demon's head."

Algol is the prototype eclipsing variable star; its brightness changes are obvious to the unaided eye if you make a point of looking for them. Every 2.87 days Algol fades from magnitude 2.1 to 3.4 during a span of about four hours, as the bright white star becomes partially eclipsed by an unseen, dimmer yellow companion. Algol remains near minimum light for nearly two more hours, then returns to its normal brightness in another four. Check it whenever you look at the constellations and eventually you'll catch it in the act. Compare it to Gamma (γ) Andromedae, magnitude 2.2, the bright foot of Andromeda labeled on our chart on page 20. Most of the time the two are about equal. But

The Double Cluster as seen through a low-power, rich-field scope. Note the orange-red giants sprinkled among the hotter white stars, especially in the eastern (left) cluster. This field is 1.7° across, wider than even the lowest-power view in most telescopes.

at minimum Algol is fainter than Epsilon (ε) Persei, magnitude 2.9. The predicted times of Algol's minima are available online at SkyandTelescope.com/observing /objects/variablestars/article_108_1.asp.

Almost halfway from Algol to Gamma Andromedae lies the nice open star cluster **M34.** It shows as a faint smudge through binoculars or a good finderscope. A 2.4-inch telescope reveals about 30 stars. Through my 4.1-inch (105-millimeter) scope magnifying 47×, I see 18 moderately bright stars and 55 fainter ones. The periphery of the cluster looks square. Its center contains most of the brighter stars, many arranged in pairs. The most eyecatching pair is the double star h1123; its two 8th-magnitude white suns mark the northwest edge of the cluster's core. Separated by 20″, they are easy to split at low power.

The brightest star of Perseus is Mirfak, Alpha (α) Persei. It's part of a huge, elongated group of stars known as the **Alpha Persei Association** or Melotte 20, easily visible with the naked eye as an enhancement of the Milky Way. The group is about 5° long, too large to fit in the view of most telescopes, but many of its stars (mostly to Mirfak's southeast) show well through a good finderscope.

The 4th-magnitude star Lambda (λ) Persei lies 7° east of Mirfak. From here we can hop 1.6° northeast to the open cluster **NGC 1528.** A 2.8-inch scope shows a rich group of faint stars about 0.4° across. In my 14 × 70 binoculars, I can

A close-up of NGC 1528. Martin Germano used an 8-inch f/5 reflector for this 30-minute exposure on hypersensitized Kodak Technical Pan 2415 film. The field is 1 ½° across.

count more than 50. Several arcs of stars reach out from the central mass, making this cluster look a little like a spider.

Southeast of our spider by 1.3° we find **NGC 1545.** It's just 0.4° east of the 4.6-magnitude star known as b Persei. NGC 1545 doesn't look much like a cluster through a small telescope, but a colorful isosceles triangle of three fairly bright stars marks its center. Shining at magnitudes 7, 8, and 9, they appear orange, yellow, and bluish white, respectively.

Let's return to Mirfak and move 8° northwest to the naked-eye star **Eta (η) Persei,** the constellation figure's northernmost star on our all-sky map. This is a beautiful double star for small telescopes. The golden primary, magnitude 3.8, has an 8.4-magnitude bluish companion 28″ to the west-northwest. This pair is an easy split at 25×.

Four degrees west-northwest of Eta, in the direction of Cassiopeia, is the stunning Perseus **Double Cluster** (Caldwell 14). In his classic *Field Book of the Skies* (1929), William Tyler Olcott called this "one of the finest clusters for a small telescope. The field is simply sown with scintillating stars, and the contrasting colors are very beautiful." As the name implies, this agglomeration consists of two star clusters side by side: NGC 869 and NGC 884. They just fit into a 1° field of view.

The western cluster, NGC 869, shows two fairly bright stars and 60 fainter ones. The dense center of the cluster contains a striking pattern of stars that begs for a game of dot-to-dot. My best effort results in a comical bat face. The cluster's brightest star marks the bat's nose; the curve of stars to its southeast is a smile. On the opposite side of the nose I can imagine bat ears. None of this shows well in photographs.

The eastern cluster, NGC 884, is a little larger than its companion, with about 80 moderately bright to very faint stars. The colors that Olcott refers to are subtle through a small telescope and mostly show up as pale orange stars in NGC 884.

The clusters are probably physically related. NGC 869 has a cataloged distance of 6,800 light-years compared to 7,600 for NGC 884. Other recent studies place them even closer together. Both lie in the so-called Perseus Arm of our galaxy, the next spiral arm outward from our own.

Now I'd like to suggest a challenge object. Although the **California Nebula (NGC 1499)** is extremely dim, it's a target best suited to binoculars or small scopes with very low powers and wide fields. The nebula is large (2.7° by 0.7°) and forms a roughly California-shaped arc about 1° north of 4th-magnitude Xi (ξ) Persei.

Slowly scan back and forth across the patch of sky

outlined in green on the map. Faint nebulosity is often easier to detect when seen in motion. Very dark skies are usually required for the California Nebula, but it responds well to a hydrogen-beta nebula filter even in suburban skies. If you can detect the California Nebula, you'll have bagged something few observers have ever seen.

A Hero's Celestial Sights

Object	Type	Mag.	Dist. (l-y)	RA	Dec.
Algol	Variable star	2.1–3.4	93	3^h 08.2^m	+40° 57′
M34	Open cluster	5.2	1,400	2^h 42.1^m	+42° 45′
Alpha Per Assn.	OB association	1.2	600	3^h 22^m	+48.6°
NGC 1528	Open cluster	6.4	2,600	4^h 15.4^m	+51° 14′
NGC 1545	Open cluster	6.2	2,600	4^h 20.9^m	+50° 15′
η Per	Double star	3.8, 8.4	1,000	2^h 50.7^m	+55° 54′
Double Cluster	Open clusters	5.3, 6.1	6,800, 7,600	2^h 21^m	+57.1°
NGC 1499	Diffuse nebula	—	2,000?	4^h 01^m	+36.5°

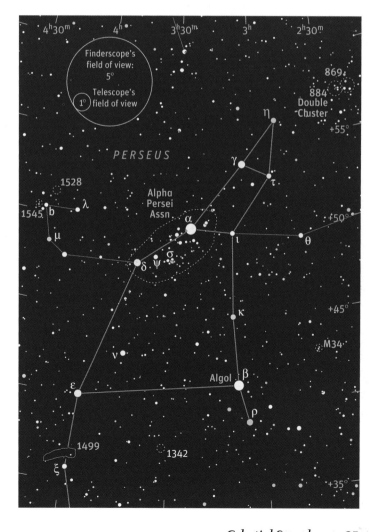

The circles in the chart show the size of a typical finderscope's field of view (5°) and a typical telescope's view with a low-power eyepiece (1°). North is up and east is left. To find which way is north in your eyepiece view, nudge your telescope slightly toward Polaris; new sky enters the view from the north edge. (If you're using a right-angle star diagonal at the eyepiece, it probably gives a mirror-image view. Take it out to see an image matching the map.)

Winter Wonders

The central rectangle of the planetary nebula M76 is what most observers see at a casual glance, and this odd shape has inspired such names as the Cork and the Little Dumbbell. William C. McLaughlin, using a 12½-inch f/9 reflector in Oregon, combined 21 CCD images to create this view.

JANUARY NIGHTS often turn bitter cold for those of us in midnorthern latitudes, but they offer wonders well worth the frigid fingers and numbed nose. The crisp, dry air often tempts us with exceptional transparency, luring us out into the star-pierced night. So bundle up and join me for a winter sky tour that will sweep us across a swath of sky from northern Perseus into the neighboring constellation Cassiopeia.

We'll begin with **M76,** a small planetary nebula that goes by such nicknames as the Little Dumbbell, Barbell, Cork, and Butterfly. To locate M76, start at 4th-magnitude, blue-white Phi (ϕ) Persei. Placing Phi at the southern edge of a low-power field, you should see a distinctly yellow-orange, 7th-magnitude star to the north. M76 is just 12′ west-northwest of this star.

Through my 4.1-inch refractor at 127×, M76 most closely suggests a cork to me. It is fairly bright and bar shaped, running northeast to southwest with a slight pinch in the middle. The nebula looks patchy, and the southwestern lobe appears to be the brighter one. M76 lies across the hypotenuse of a right triangle formed by three very faint field stars. If you are battling light pollution, an oxygen III or narrowband filter can help.

Under the dark skies of the northern Adirondack Mountains in New York, I can just glimpse what seems to be a dull zone separating the two lobes of the nebula. Its double nature earned M76 two separate designations in J. L. E. Dreyer's *New General Catalogue of Nebulae*

and Clusters of Stars (London, 1888). The southwestern section is NGC 650, while the northeastern section is NGC 651. I can sometimes descry vague hints of the diaphanous wings that loop outward from the long sides of the nebula. In long-exposure images, these wings do make M76 look very much like a butterfly.

The fabulous **Double Cluster** is our next stop, a naked-eye wonder known since antiquity. The star chart of page 20 can help in finding it. Look for the two stars labeled Gamma (γ) and Delta (δ) in **W**-shaped Cassiopeia. Imagine a line from the first through the second, continue farther for twice that distance, and you'll be led right to the Double Cluster. It's a hazy patch embedded in the dim band of the Milky Way.

Although the pair can be crammed into a 1° field of view, it is better appreciated with a short-focal-length telescope giving fields 1.5° wide or more. Having ample dark sky around the clusters helps to offset their sparkling beauty. These clusters (also known as Caldwell 14) form a true pair, for they are known to have approximately the same age and distance from us.

The eastern member of the Double Cluster is **NGC 884.** With my 4.1-inch scope at 68×, I see about 80 bright to very faint stars within this group. Two small knots of bright stars lie close together, southwest of center. NGC 884 is flecked with a handful of orange stars, two of which seem orphaned between the clusters. The orange stars are variables, and through a small telescope it may be easier to perceive their colors when they are near maximum light.

The western cluster is designated **NGC 869.** This one appears a shade smaller than its companion and more highly concentrated. It has two fairly bright stars, and I count about 60 fainter ones. The central bright star has a distinctive bowl of stars to its southeast. With his 4-inch refractor, California amateur Ron Bhanukitsiri sees this as a glowing eye and eyebrow.

After the Double Cluster, let's take stock — **Stock 2,** to be precise. A curving chain of 6th- to 8th-magnitude stars winds 2° northward from NGC 869 to Stock 2. Together, the Double Cluster and Stock 2 are a compelling sight through 50-mm binoculars. Be sure to use a low-power, wide-field eyepiece for telescopic

The large nebulosity called IC 1805 is at top center in this wide-field scene, which also includes the spectacular Double Cluster (lower right). Jon A. Kolb employed a Borg 100ED f/4 refractor and Kodak Ektachrome E200 film at Badger Spring, Colorado.

views, since Stock 2 is about 1° across.

My little refractor at 47× shows several dozen stars in a loosely scattered group. The brightest stars make a stick-figure man with his head to the west and spread legs to the east. His arms are upraised and curved as though showing off his muscles. For this reason Massachusetts amateur John Davis has dubbed Stock 2 the Muscle Man Cluster, and the name seems very apt.

Next we'll move 3° northeast of Stock 2 to the nebulous cluster **IC 1805.** At 68×, it's a coarse group of 40 stars about 13′ across. A circlet of stars containing the *lucida* of the group (an old term for "brightest member") sits at the heart of the cluster with arms of stars of 8th magnitude and fainter radiating from it. The central star is the double **Stein 368** and shows a faint companion 10″ to the east.

Dropping the magnification to 17× gives my scope a 3.6° field of view, and I can see extensive nebulosity in and around the cluster. Using an oxygen III filter makes it much more apparent. The brightest areas include the cluster itself and a wide patch running east and then curving north of the cluster. Starting at the eastern side of the wide patch, a fainter loop curves south of the cluster and then up its west side. Both loops are very patchy and together span about 1½°. A small, bright, detached portion, **NGC 896,** lies 1° northwest of the cluster. Under dark skies, this large complex has been seen in instruments as small as 30-mm binoculars.

Just 1.2° east of IC 1805 is another cluster, **NGC 1027.** At 87×, I see one 7th-magnitude star loosely surrounded by approximately 40 faint to extremely faint stars within 17′. The cluster's brighter members appear to spiral out from the central star for 1½ turns.

An interesting knot of stars lies a little under 1° south-southwest of IC 1805. At 87×, **Markarian 6** is not an obvious cluster, but it has a noteworthy shape. Look for four stars of magnitude 8.5 to 9.7 in a slightly curved line running nearly north-south. Five dimmer stars join the one at the southern end to make an arrowhead, while the three to the north form the arrow's tail.

I've seen so many enchanting sights in this part of the sky that I'm glad I have space in this book to share them with you. Turn to pages 150, 152, and 154 to discover more treasures in Cassiopeia's domain. Part of the beauty of observing is in knowing that there's always more to see.

Clusters and Nebulae Where Perseus and Cassiopeia Meet

Object	Type	Mag.	Size/Sep.	Dist. (l-y)	RA	Dec.
M76	Planetary nebula	10.1	1.7′	4,000	1ʰ 42.3ᵐ	+51° 35′
NGC 884	Open cluster	6.1	30′	7,600	2ʰ 22.3ᵐ	+57° 08′
NGC 869	Open cluster	5.3	30′	6,800	2ʰ 19.1ᵐ	+57° 08′
Stock 2	Open cluster	4.4	60′	1,000	2ʰ 15.6ᵐ	+59° 32′
IC 1805	Open cluster	6.5	20′	6,000	2ʰ 32.7ᵐ	+61° 27′
IC 1805	Emission nebula	—	96′ × 80′	6,000	2ʰ 32.8ᵐ	+60° 30′
Stein 368	Double star	8.0, 10.1	10″	6,000	2ʰ 32.7ᵐ	+61° 27′
NGC 896	Emission nebula	7.5	20′	6,000	2ʰ 24.8ᵐ	+62° 01′
NGC 1027	Open cluster	6.7	20′	3,000	2ʰ 42.6ᵐ	+61° 36′
Mrk 6	Open cluster	7.1	6′	2,000	2ʰ 29.7ᵐ	+60° 41′

M76's size refers to the length of this nebula's distinctive, bright bar.

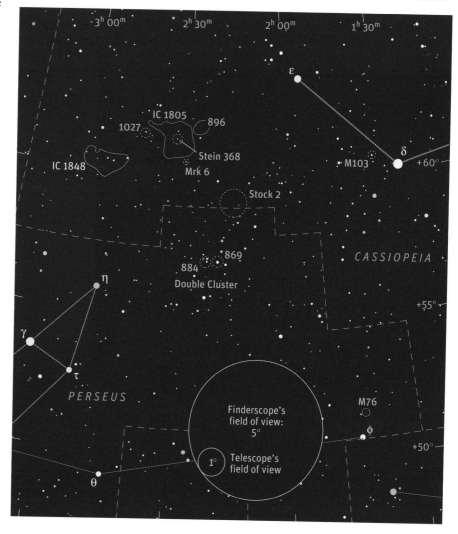

January's Celestial Menagerie

THIS TIME we're going to tour the patch of sky around the spot where the constellations Pisces, Cetus, and Aries meet. We'll start in Pisces, the Fishes, at the star marked Eta (η) on the finder chart on the opposite page. A distinctly yellow star, it resides in the ribbon that ties the two fishes of Pisces together.

Moving 1.3° east-northeast of Eta Piscium will bring us to the galaxy **M74.** With a visual magnitude of about 9.4, M74 would seem to be an easy target, but it's not. The reason is that 9.4 is the galaxy's total (integrated) magnitude, which is how bright M74 would look if its light were gathered into a starlike point. Since its glow is actually spread over a fairly large area, M74 has a very low surface brightness, making it one of the toughest Messier objects to spot; so much so that it has been nicknamed the Phantom Galaxy.

In 14 × 70 binoculars M74 appears as a moderately large galaxy. It is round and extremely faint with a uniform surface brightness. In my 4.1-inch (105-millimeter) refractor at 17×, it fits in the same field with Eta Piscium. Increasing the magnification to 87×, I see M74

Celestial smirk? At least that's one visual interpretation of this novel juxtaposition of two 7th-magnitude stars and the nearly edge-on spiral galaxy NGC 1055. Kim Zussman made this photograph with a 14½-inch classical Cassegrain reflector.

as roundish and about 7' or 8' across with several very dim stars superposed on the galaxy's disk. It has a faint halo with a large core that grows weakly brighter toward the center. A little mottling can be glimpsed.

If you have trouble locating this elusive galaxy, here are some things you can try. With a low-power eyepiece, center your view on the 7th-magnitude orange star that lies ¼° north-northeast of Eta. If your telescope has a drive, turn it off, and after five minutes M74 will be in the eyepiece just a bit north of center. Watching carefully when you expect the galaxy to drift into the field will make it easier to discern its ghostly glow contrasting with the darker background sky. Once you think you've spotted M74, try lightly tapping the side of your scope. A dim object in motion is often easier for the eye to detect than a stationary one. Draping a dark cloth over your head to block stray light can also help.

Averted vision will greatly improve the view. Look to one side of the galaxy instead of directly at it. Turning your gaze so that the object appears between your nose and the direction you're looking usually works best.

Now let's visit a galaxy that is easier to see than M74. It is located in the head of Cetus, the Whale. The loose ring of naked-eye stars in the constellation's northeastern corner will serve as our guidepost. Looking at the finder chart opposite, you'll see that the southernmost star in this asterism is Gamma (γ) Ceti. It is joined by a constellation line to Delta (δ) Ceti, the blue-white star to its south. The galaxy **M77** is 52' east-southeast of Delta; they share the same low-power field in a small telescope.

Through 14 × 70 binoculars, I see M77 as fairly faint,

Although the face-on spiral galaxy M74 appears stunning in color CCD images such as this one by Robert Gendler, it is a challenging object to see in small telescopes because of its low surface brightness. See the text for several tips on how to hunt down elusive targets.

small, and round with a dim field star nearly touching the edge. In the 4.1-inch scope at 87×, its gauzy halo is slightly oval and about 2′ long running north-northeast to south-southwest. M77 shows a very small, bright core with a nearly stellar nucleus. The dim star just mentioned, of magnitude 10.8, is comparable in brightness to the nucleus and lies near the galaxy's east-southeast rim. At low power, a casual glance might even trick you into thinking you're seeing a double star. M77 has a total visual magnitude of 8.9 and a higher surface brightness than M74. The brightness distribution is also more uneven than that of M74.

M77 is one of only two Seyfert galaxies in the Messier catalog, the other being M106 in Canes Venatci. A Seyfert galaxy has a brilliant nucleus and a spectrum that suggests immense gas clouds are rapidly moving out of its core. The enormous energy required to generate this velocity originates in the galaxy's center, where a supermassive object (probably a black hole) devours surrounding material, turning the galactic nucleus into a miniature quasar. The central object within M77 is estimated to be about 10 million times more massive than our Sun.

M77 forms a gravitationally related pair with the adjacent galaxy **NGC 1055.** By moving either 0.6° north-northwest of M77 or the same distance east-northeast of Delta Ceti, you will come to a pair of 7th- and 8th-magnitude stars 7′ apart and aligned east-west. NGC 1055 is just 6′ south of this pair. Through my refractor at 127×, this tenuous, spindle-shaped galaxy appears about 1′ wide and 4′ long running east-southeast to west-northwest. An 11th-magnitude star lies close along the north side. Although NGC 1055 has been seen in scopes with apertures as small as 2.4 inches, it is a challenging target unless the sky is moderately dark. The total visual magnitude of NGC 1055 is 10.6, and its surface brightness is between that of M77 and M74.

We end this tour with a star-hop to an unfamiliar open star cluster in Aries, the Ram, with a stop at a pretty multiple star. Our starting point is 4.3-magnitude Mu (μ) Ceti, the northernmost star in the Cetus ringlet. Use your finder or a low-power eyepiece to sweep 2.3° north to 5.2-magnitude 38 Arietis, then continue another 2.9° to 5.8-magnitude Omicron (o). From this star you can move 2.4° north-northeast to the blue-white, 5.3-magnitude star **Pi (π) Arietis.** At 87×, Pi is a triple system with a close, fairly bright, yellow-white companion and a dim companion about eight times farther away. Both lie approximately east-southeast, but not quite in a straight line. The close pair is comfortably split at 127×.

Finally, our little-known cluster, **Dolidze-Džimšelejšvili 1,** is ½° west-southwest of Pi and easily shares the same low-power field. My

4.1-inch scope at 87× shows six moderately bright stars (magnitudes 8.5 to 10.9) and five 13th-magnitude stars in an area about 9′ across. Although this is a scanty group, it stands out well against the region's star-poor background. Those with larger scopes might enjoy trying to sort out the colors of the six brightest stars. Their color indexes indicate that two are white, two are yellow, and two are orange.

On some clear January evening, aim your telescope toward the celestial shore where the Fishes, the Whale, and the Ram have gathered and discover the astronomical treasures they are showing off to each other.

Where Pisces, Cetus, and Aries Meet

Object	Type	Mag.	Size/Sep.	Dist. (l-y)	RA	Dec.
M74	Galaxy	9.4	10.5′ × 9.5′	32 million	01ʰ 36.7ᵐ	+15° 47′
M77	Galaxy	8.9	7.1′ × 6.0′	47 million	02ʰ 42.7ᵐ	−00° 01′
NGC 1055	Galaxy	10.6	7.6′ × 2.7′	41 million	02ʰ 41.7ᵐ	+00° 27′
Do-Dž1	Open cluster	—	12′	—	02ʰ 47.5ᵐ	+17° 15′
π Ari	Multiple star	5.8, 8.0, 10.7	3.3″, 25.1″	600	02ʰ 49.3ᵐ	+17° 28′

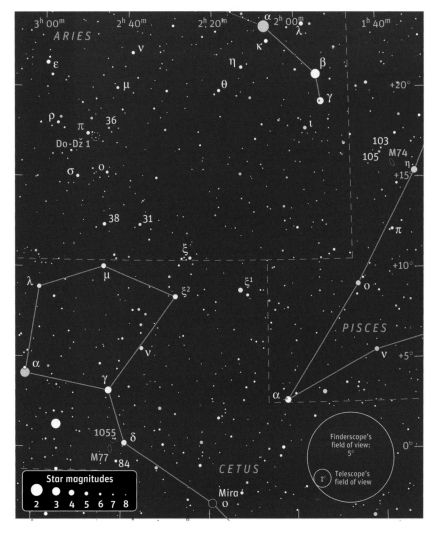

Northern Nights

IT OFTEN SEEMS that we eagerly await seasonal sights appearing in the east while taking for granted those that stay with us all night and all year long. Such is the fate of Camelopardalis, the Giraffe, which is circumpolar for midnorthern latitudes.

Camelopardalis is a "modern" constellation whose origin dates back just to the early 17th century. Its creation is generally attributed to Dutch cartographer Petrus Plancius, who used it to fill an oddly shaped region of faint stars that had not been assigned to any constellation. Gazing northward, we see no clear outline of the Giraffe — only his spots scattered across the sky.

In this essay I'll explore the area around southwestern Camelopardalis, where the hindquarters of the Giraffe meet the queenly toes of neighboring Cassiopeia. Small-scope enthusiasts will find it rich in deep-sky wonders large and small, effortless and challenging.

Let's start with **Stock 23,** an open cluster that straddles the official boundary between these two constellations. Stock 23 is sometimes called Pazmino's Cluster after New York amateur John Pazmino, who chanced upon the group with a friend's 4.3-inch refractor while sweeping for the Double Cluster in Perseus. His serendipitous sighting was brought to the attention of amateur observers by Walter Scott Houston in his Deep-Sky Wonders column in the March 1978 issue of *Sky & Telescope.*

A sparkling star cluster within a much larger nebula, IC 1848 glows with the characteristic red hue of ionized hydrogen. Working near Houston, Texas, Bobby Middleton made this composite of two hour-long exposures with a 10-inch Takahashi telescope. Since human eyes are largely colorblind with faint diffuse objects, most people see the nebula as a soft, grayish glow. This view is 2⅓° across.

Stock 23 is visible through binoculars as a little group of four stars in an irregular trapezoid resembling Draco's head. With my 4.1-inch refractor at 87×, I count 27 stars of magnitude 7.5 and fainter, spread across 13′. Most are arrayed in an oval outline running northwest to southeast with two extensions springing from the southwestern side. The unusual shape begs for games of dot-to-dot. To me it looks like the face of a lop-eared rabbit with ears at half-mast. The base of one ear is a nearly matched close double star, Σ362.

Stock 23 is involved in nebulosity that I have never been able to detect. While scanning the area, however, I've noticed a rather bright nebula 2° west-northwest. **IC 1848** appears about 1½° long and ¾° wide and is best seen with a low-power, wide-angle eyepiece. It is visible through my little refractor at 17× and 28× even without a filter. A narrowband light-pollution filter improves the view a bit, while both oxygen III and hydrogen-beta filters do better still. Your results may vary depending on the size of your scope and the darkness of your sky. The nebula is patchy and brightest in wide bands along the north, east, and west sides.

The nebula enshrouds two open clusters, neither particularly obvious. The western one shares the designation **IC 1848** with the nebula and surrounds the widely spaced bright double star Σ306 AG. At 68× I see many very faint stars scattered across 18′ with the densest areas lying south and east. To the east, **Collinder 34** is larger and centered on a pair of stars (magnitudes 8 and 9) with radial strings of stars branching outward from it.

Seen nearly face on, the sprawling spiral arms of the galaxy IC 342 in Camelopardalis span an area of sky almost as large as the Moon. Fine details of their structure stand out well in this image by Robert Gendler of Avon, Connecticut. He used a 12½-inch telescope and an SBIG ST-10E camera.

The cluster **Trumpler 3,** found 3.2° north of Stock 23, is a little more obvious. At 87× I see a glimmering splash of 35 stars, 9th magnitude and fainter, in a loose swarm with indefinite borders. Three of the brightest stars form a north-south line west of center; a fourth lies in the eastern edge of the cluster.

If you scan 5° east from Trumpler 3, you'll come to one of the most noteworthy asterisms in the sky. The late Lucian Kemble of Canada stumbled across this group by pure accident while scanning the sky with 7 × 35 binoculars. He called it a "beautiful cascade of faint stars tumbling from the northwest down to the open cluster NGC 1502." Kemble sent his description and a drawing to Walter Scott Houston, who published them in his December 1980 Deep-Sky Wonders column. In later columns, Houston referred to this charming alignment of stars as **Kemble's Cascade,** and the name stuck.

Kemble's Cascade is a 2½° line of 7th- to 9th-magnitude stars. One 5th-magnitude star punctuates the middle of the chain. In my little refractor, a wide-angle eyepiece yielding 17× and a 3.6° field gives a beautiful view, revealing about 20 stars. Near the southeastern end of the Cascade, **NGC 1502** is a pretty cluster with many faint stars crowded around a pair of yellow-white, 7th-magnitude suns (Σ485). At 68× I count 25 stars gathered into a squat triangle.

Dropping 1.4° south of NGC 1502, we find the little planetary nebula **NGC 1501.** It appears small, round, and fairly bright through my 4.1-inch scope at 68×. This planetary takes on a little more character with each increase in aperture. Features to look for include a somewhat dark center, brighter north-eastern and southwestern edges, a faint central star, and a slightly oval shape.

Our final stop will be **IC 342,** or Caldwell 5, one of the closest galaxies beyond our own Local Group. Starting at the northwestern end of Kemble's Cascade, move 1.8° north to a 4th-magnitude orange star with a 7th-magnitude yellow star 21' east. North 1.6° and a little west of the orange star, you'll find a similarly spaced pair consisting of a 6th-magnitude white star and a 7th-magnitude deep-yellow star. The parallelogram formed by the four should be easily visible in a finder. IC 342 is centered 54' north of the white star. Placing this star at the southern edge of a low-power field will put the galaxy near the northern edge.

For a good view of this large, low-surface-brightness galaxy, be sure to center it in your field of view. In my 4.1-inch scope at 28×, this pretty galaxy is a vaporous phantom spangled with faint stars. It appears oval, its long dimension running north and south with a length of 12'. From a dark-sky site with his 4-inch refractor, *Sky & Telescope* contributing editor

and noted observer Stephen James O'Meara has been able to trace out IC 342's three main spiral arms.

IC 342 is a member of the Maffei 1 Group of galaxies. It is relatively nearby at 11 million light-years, rivals our own galaxy in luminosity, and is one of the brightest galaxies in the northern sky. Its diminished glory as seen through telescopes is largely due to our vantage point. IC 342 lies only 10.6° above the galactic plane of the Milky Way, where it is highly obscured by intervening clouds of gas and dust.

Far-North Sights for a January Evening

Object	Type	Mag.	Size	Dist. (l-y)	RA	Dec.
Stock 23	Open cluster	5.6	14′	—	3ʰ 16.3ᵐ	+60° 02′
IC 1848	Emission nebula	7.0	100′ × 50′	6,500	2ʰ 53.5ᵐ	+60° 24′
IC 1848	Open cluster	6.5	18′	6,500	2ʰ 51.2ᵐ	+60° 24′
Cr 34	Open cluster	6.8	24′	6,500	2ʰ 59.4ᵐ	+60° 34′
Tr 3	Open cluster	7.0	23′	—	3ʰ 12.0ᵐ	+63° 11′
Kemble 1	Asterism	5 to 9	150′	—	3ʰ 57.4ᵐ	+63° 04′
NGC 1502	Open cluster	5.7	7′	2,700	4ʰ 07.8ᵐ	+62° 20′
NGC 1501	Planetary nebula	11.5	52″	4,200	4ʰ 07.0ᵐ	+60° 55′
IC 342	Spiral galaxy	8.3	21′	11 million	3ʰ 46.8ᵐ	+68° 06′

Angular sizes are from catalogs or photographs; most objects appear somewhat smaller when a telescope is used visually.

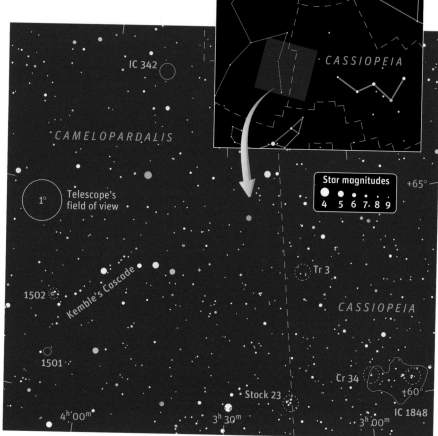

Fireflies

Many a night I saw the Pleiads,
rising thro' the mellow shade,
Glitter like a swarm of fire-flies
tangled in a silver braid.

— ALFRED, LORD TENNYSON, *Locksley Hall*

HIGH IN THE SOUTH on crisp January evenings, we see the magnificent **Pleiades (M45)** sparkling in icy splendor. This spectacular star cluster is easily visible to the unaided eye and has been known since antiquity. It is commonly known from Greek mythology as the Seven Sisters and may be the most widely recognized star pattern in the northern sky after the Big Dipper. This remarkable group has features to thrill skygazers with the most modest equipment and to challenge those with the largest.

Six stars gathered into the shape of a tiny dipper can be seen in suburban skies. Atlas, the father of the Pleiades, marks the handle while his daughters Alcyone, Merope, Electra, Maia, and Taygeta form the bowl.

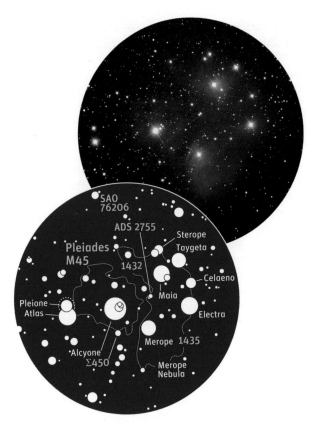

The Pleiades glitter in the evening sky from October to mid-April. From a dark-sky site, you can easily see all the brightest members with your naked eye. Through a small telescope, and with averted vision (staring to one side of the eyepiece field), this open cluster appears as a swarm of stars.

Darker skies and keen eyesight can reveal a few dimmer stars. Very close to Atlas on the north is his wife, Pleione; Sterope is north of her sister, Maia; and Celaeno huddles between Taygeta and Electra.

Many stars with no common name also belong to the Pleiades star cluster, and some may be glimpsed with the unaided eye. I can often see 10 stars from my semirural home in upstate New York, and the noted observer Walter Scott Houston could pick out 18 under exceptional skies. A 2.4-inch (60-millimeter) scope can swell the star count to more than 60, and a 6-inch instrument triples that number.

In some ways a small telescope is best for viewing the Pleiades. One with a short focal length can fit this large group within a single field of view, since the main mass of stars is about 1.6° across. If your lowest-power eyepiece can not encompass the entire cluster, be sure to make your first observations through the telescope's finder. Examine the stars and use the view of the brighter ones to help increase your naked-eye count.

A small telescope is also the best way to glimpse the delicate nebulosity that enfolds the Pleiades. Contrary to what some sources claim, this nebulosity is not the tattered remains of the cloud from which the Pleiades were born. Cloud and cluster are moving in different directions through space and were not together when the stars were formed, yet this coincidental meeting makes for a sight of delicate beauty.

Seen in this image by Bob and Janice Fera, the Pleiad Merope is embedded in a bluish, wispy-looking cloud of gas and dust. This Merope Nebula (NGC 1435) is part of a larger nebula called the Taurus Dark Cloud Complex.

It is easy to see nebulosity around the bright Pleiads. Just breathe on your eyepiece on a wintry night and notice the effect you get. To be sure what you see is real, concentrate first on Merope. The nebula is brightest and largest around this star, spreading mostly to its south. You'll be looking for a haze only slightly brighter than the background sky, so use a low power to capture more of the sky around it for comparison. This diaphanous cloud is known as the **Merope Nebula (NGC 1435).**

Now look for nebulosity surrounding the other bright stars in the bowl of the Pleiades' tiny dipper shape. The **Maia Nebula (NGC 1432),** surrounding its namesake star, may be the next-easiest patch to detect.

Finally, examine Atlas and Pleione. If you see nebulosity here, your optics have probably fogged over. The small amount of mist surrounding these stars is unlikely to be detected. As a check, swing your scope over to similarly bright stars in the nearby Hyades (shown on our all-sky map on page 20). These should not display any nebulosity.

M45 is a relatively young star cluster about 100 million years old. As such, it has many hot, blue-white stars. Still, hints of other colors can be seen within the cluster. A pair of 8th-magnitude stars **(ADS 2755)** in the dipper's bowl is easily split at 25× and shows nicely contrasting white and orange hues. A 6th-magnitude star **(SAO 76206)** at the extreme northern edge of the group also has an orange tint. An attractive, bent chain of six stars leads south from Alcyone and then turns south-southeast. The last two stars in the chain are orange, but the color of the end star is difficult to detect through a small scope.

The star nearest to Alcyone in this chain is the close double **Σ450.** It consists of a 7.3-magnitude primary with a 9.3-magnitude secondary 6.1″ to the west. Try powers of around 80× to split the pair.

Many other multiple stars inhabit the Pleiades. Sterope is actually a pair of 6th-magnitude, blue-white stars just a little too close to be separated with the unaided eye. Taygeta is also a wide double with an 8.1-magnitude secondary 69″ north-northwest of the 4.3-magnitude primary. The pair is well split at 17× in my 4.1-inch refractor. Bright Alcyone has a little triangle of 6.3- to 8.7-magnitude stars to her north-northwest. The brighter two are white, while the dimmest has a yellowish cast.

Atlas (magnitude 3.6) and Pleione (magnitude 5.1) form a nice naked-eye double under dark skies and are otherwise easy targets with only the slightest optical aid. Pleione spins so rapidly that it is squashed into an oblate spheroid. This breakneck rotation causes the star to throw off a gaseous shell of material at irregular intervals. When this happens, Pleione first brightens

and then dims below normal before returning to its usual magnitude. The last period of such variability ran from 1972 to 1987 with a magnitude range of 4.8 to 5.5. Many cultures have stories of a lost Pleiad, and only six of the seven stars of lore are obvious at a casual glance. The variability of Pleione led astronomer Edward C. Pickering to propose that Pleione may once have shone with a greater brilliance, giving rise to these legends from around the world.

Exploring the Pleiades

Object	Type	Mag.	Dist. (l-y)	RA	Dec.
Pleiades (M45)	Open cluster	1.2	400	03ʰ 47.0ᵐ	+24° 07′
NGC 1435	Reflection nebula	4.2	400	03ʰ 46.1ᵐ	+23° 47′
NGC 1432	Reflection nebula	3.8	400	03ʰ 45.8ᵐ	+24° 22′
ADS 2755	Double star	8.3, 8.4	400	03ʰ 46.3ᵐ	+24° 11′
SAO 76206	Star	6.4	700	03ʰ 48.1ᵐ	+24° 59′
Σ450	Double star	7.3, 9.3	600	03ʰ 47.4ᵐ	+23° 55′

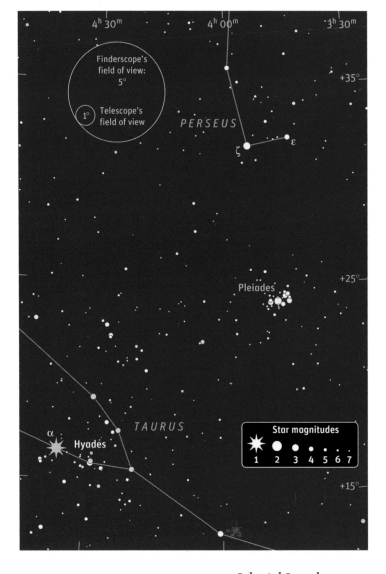

On this chart north is up and east is left. The circles show fields of view for a typical finderscope (5°) or a small telescope with a low-power eyepiece (1°). To find north through your eyepiece, nudge your scope toward Polaris; new sky enters the view from the north edge. (If you're using a right-angle star diagonal, it probably gives a mirror image. Take it out to see an image matching the map.)

Looking the Bull in the Face

THE DISTINCTIVE **V**-shaped face of Taurus, the Bull, is now high in the evening sky, with the the Bull's glaring eye marked by the bright star Aldebaran. The dimmer stars of the Bull's face belong to the **Hyades,** one of the nearest open clusters about 150 light-years from us. Orange-hued Aldebaran is a foreground star only 65 light-years away. This red giant is about 40 times larger across than our Sun and 150 times more luminous.

Even small telescopes can't embrace the entire Hyades Cluster at once, but the group's striking colors and double stars make it well worth exploring. For example, let's visit the little triangle of stars 1½° west of Aldebaran. The entire triangle will fit within a low-power field, each corner anchored by a widely separated pair of stars. The brightest is the naked-eye double star **Σ I 10.** It is composed of the stars Theta¹ (θ¹) and Theta² (θ²) Tauri. Through my 4.1-inch refractor, 3.4-magnitude Theta² appears white, while slightly dimmer 3.9-magnitude Theta¹ is golden.

The eastern point of the triangle is occupied by **LDS 2246.** The 4.8-magnitude primary has a 6.5-magnitude companion to the southeast. They are spaciously separated by 250″ and can be split even with binoculars. Through my little scope they show a subtle color contrast, the brighter star appearing white and the other more of a yellowish white. An unrelated pair of stars marks the triangle's northern point. The 5th-magnitude, deep-yellow star **75 Tauri** has an orange attendant of 8th magnitude lying 228″ to the south-southwest.

Don't expect to see a Crab Nebula this colorful in your small scope — or, in fact, any telescope. The human eye is almost colorblind to dim light and, in addition, responds poorly to the long wavelengths of the Crab's flamelike protuberances. But the nebula should resemble the soft, underlying blue-gray cloud seen in this composite image by Russell Croman.

Near the center of the triangle, a solitary 6.6-magnitude star appears pale yellow, and between this and LDS 2246 is an 8.7-magnitude star that looks orange.

Leaving the Hyades and moving 3½° northeast of Aldebaran, we come to the large open cluster **NGC 1647.** Through my 4.1-inch refractor at 47×, I count 75 stars from 9th to 12th magnitude in ½°. The cluster appears very roughly rectangular, with extensions at the northeastern and southwestern corners. The double star **AG 311** can be seen a few arcminutes north of center; it consists of two 9th-magnitude suns. Two orange stars (6th and 7th magnitude) stand just off the cluster's southern edge.

Another large, open cluster dwells 2° north of the 4.6-magnitude star Iota (ι) Tauri. If your sky is too bright to see Iota, you can locate **NGC 1746** by noting that it lies three-fifths of the way from Aldebaran to the tip of the Bull's northern horn, Beta (β) Tauri.

Through my 4.1-inch scope at 17×, I see 60 moderately bright to faint stars in an irregular, loose group about ⅔° across. The borders of the cluster seem ill defined, especially in the north. The four brightest stars cradle its east and southeast edges. In my mirror-imaged view, they form an **L** with a curved back. A concentration of stars, faint to extremely faint, is nestled in the crook of the **L.**

Although this cluster is labeled NGC 1746 on most star atlases, its identity is not completely clear. Some atlases plot two or even three overlapping clusters here, the others being **NGC 1750** and **NGC 1758.**

William Herschel discovered NGC 1750 and NGC 1758 on the same night two centuries ago. His description for 1750 says "a Cl. of v. co. sc. Lst. join. to VII.21" (a cluster of very coarse scattered large stars joined to NGC 1758). For NGC 1758 he has "a Cl. of p. com. st. with many eS. st. mixed with them" (a cluster of pretty compressed stars with many extremely small stars mixed with them). Both correspond fairly well to my visual impressions.

What came to be called **NGC 1750** was actually not recorded until years later, by Heinrich d'Arrest, who described it as a poor cluster. He put its center just 9′ north and a little west of that given by Herschel for NGC 1750. But since the latter is quite large and irregular, d'Arrest and Herschel were probably observing the same object. Nevertheless, some observers claim that three concentrations of stars can be seen in this area. Does this look like one, two, or three clusters to you?

Recent studies of this region are promoting the idea that there are two distinct clusters here, lying at different distances but overlapping on the sky. They refer to the large and scattered group of bright stars as NGC

1750, and to the smaller concentration of fainter stars in its eastern side as NGC 1758.

Now let's take the Bull by the horns — or, at least, one of them. The tip of the Bull's southern horn is marked by blue-white Zeta (ζ) Tauri. From here, scan 1.1° northwest to find the **Crab Nebula,** a supernova remnant (SNR) and the first object in Charles Messier's catalog. M1 consists of material that was ejected from a supernova explosion seen in AD 1054. More than 900 years later, my little scope at 87× shows an oval nebula 5′ long by 3′ wide. The still-expanding glow stretches northwest to southeast, narrowing at one end. It looks fairly uniform in brightness across most of its face, dimming across the narrow portion and around the fringe.

John Bevis, an English physician and amateur astronomer, discovered this nebula in 1731. Messier found it independently while observing the Comet of 1758, and it's what inspired him to start his famous list of deep-sky objects. He wrote, "This nebula had such a resemblance to a comet, in its form and brightness, that I endeavored to find others, so that astronomers would not confuse these same nebulae with comets just beginning to shine."

Taurus's Deep-Sky Bounty

Object	Type	Mag.	Size/Sep.	Dist. (l-y)	RA	Dec.
Hyades	Open cluster	0.5	5.5°	150	4ʰ 27.0ᵐ	+15° 50′
Σ I 10	Optical double	3.4, 3.9	347″	149, 158	4ʰ 28.7ᵐ	+15° 52′
LDS 2246	Optical double	4.8, 6.5	250″	145, 134	4ʰ 30.6ᵐ	+16° 12′
75 Tauri	Optical double	5.0, 8.2	228″	190, —	4ʰ 28.4ᵐ	+16° 22′
NGC 1647	Open cluster	6.4	40′	1,800	4ʰ 45.9ᵐ	+19° 08′
AG 311	Double star	8.9, 9.3	33″	1,800	4ʰ 45.9ᵐ	+19° 11′
NGC 1746*	Open cluster	—	—	—	5ʰ 03.6ᵐ	+23° 48′
NGC 1750	Open cluster	6.1	40′	2,100	5ʰ 03.8ᵐ	+23° 39′
NGC 1758	Open cluster	—	20′	2,500	5ʰ 04.4ᵐ	+23° 47′
Crab Nebula (M1)	SNR	8.4	6′ × 4′	6,300	5ʰ 34.5ᵐ	+22° 01′

*NGC 1746 is presumably the same cluster as NGC 1750, while NGC 1758 is part of NGC 1750 (see text).

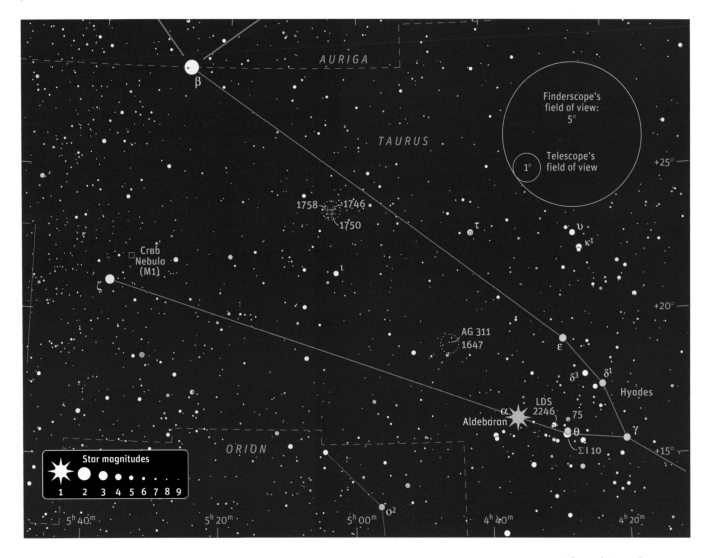

Touring Orion's Sword

THE WONDERFULLY BRIGHT constellation Orion, the Hunter, strides high across the southern sky in the depths of winter. Our evening sky map on page 21 shows Orion above the "Facing South" label and standing athwart the celestial equator. The bright stars Betelgeuse and Bellatrix mark Orion's shoulders, while Rigel and Saiph mark his legs or feet. In the middle is Orion's Belt, the eye-catching diagonal row of three stars.

From rural or suburban locations, you can see what seems to be a line of three or four faint stars dangling below Orion's Belt. This is Orion's Sword. It is encrusted with a fascinating complex of nebulae and star clusters waiting to be explored with a small telescope.

The central star of Orion's Sword may seem unusual even to the unaided eye. If your night is very dark and your vision sharp, a close look reveals that this star appears slightly fuzzy. A glance through binoculars or a good finderscope confirms the suspicion. A look through a telescope reveals the famous Great Orion Nebula, **M42,** a vast stellar nursery of glowing gas and dark dust lit by the young stars within. At first you may see only a dim, fan-shaped glow enveloping a few stars. But examine, study, and be patient. Focus your attention on different aspects, return as a friend to visit it on other nights, and you will learn to see the many faces of this intricate spectacle.

First try a low-power, wide-field view. You'll see that the brightest portion of the nebula surrounds the brightest star within it. Fainter arcs of nebulosity extend northwest and south-southeast to enclose an even fainter region glowing between them. Through my 4.1-inch telescope well away from light pollution, I can see these curving arms meet to form a closed loop about $^2/_3$° wide that sideswipes bright Iota (ι) Orionis and its associated stars to the south. This is definitely a sighting worth investing some observing practice to achieve!

Now try a medium power of about 40× to 80×. The brightest area of the nebula takes on a mottled appearance, full of swirling detail that can appear more three-dimensional than in any photograph. Many people can see the distinct greenish hue of this region, which is quite unlike the pink or red that nebulae usually display in photographs. Our eyes are less sensitive than color film to a nebula's strong, deep-red hydrogen emission but respond well to the green emission from a nebula's doubly ionized oxygen. Notice the dark protrusion of interstellar dust jutting into the bright part of the nebula from the northeast. This dark nebula is known as the Fish's Mouth. At powers of 125× and up the mottling of the bright area becomes more obvious. It has been likened to a mackerel sky.

At medium and high powers the brightest star stands clearly exposed for the multiple system it is. This is Theta¹ (θ¹) Orionis, the famous **Trapezium.** Even a 2.4-inch (60-millimeter) scope will reveal its four brightest components arranged in the shape of a tiny trapezoid (though in a telescope so small, one star appears very faint). The four are designated A, B, C, and D, from west to east: C is the brightest; B is the faintest.

With a larger scope at high power, you may be able

NGC 1981

NGC 1977

M43

M42

ι

Σ745

Σ747

Orion's Sword from top to bottom. This field is 2° tall, with north up and east left. The bright heart of M42, the Great Orion Nebula, is overexposed; for details within it see the higher-power view at right.

to see two more members of the Trapezium with some difficulty. The atmospheric seeing must be very steady so the stars focus down to sharp points. With a 3.5-inch telescope you may be able to spot star E; it's more or less between A and B as indicated in the diagram at lower right. Star F is more difficult and requires at least a 4-inch or larger scope, since it is easily lost in the glare of C. On a night of mediocre seeing no telescope will reveal it at all.

We're not finished with the Orion Nebula yet. A 7th-magnitude star lies just north of the Fish's Mouth. In a dark sky a 2.4-inch scope will show a small, roundish glow surrounding it, slightly off center. Through my 4.1-inch refractor the nebula, **M43,** takes on the "comma" shape familiar from photographs. The tail of the comma, to the northeast, appears fainter than its large dot.

Move just $\frac{1}{2}°$ north and you come to **NGC 1977,** a large, sparse bunch of stars enveloped in very faint nebulosity. My 4.1-inch shows two 5th-magnitude stars lined up nearly east-west and 15 fainter ones scattered around. The two brightest display a slight color contrast. The western star, 42 Orionis, shows the blue-white color common to many of the massive young stars in Orion. The eastern one, 45 Orionis, is yellow-white.

In a dark sky I can see that the whole group is embedded in a dim, elongated glow showing some internal structure. The brightest part of this reflection nebula curves through the two brightest stars and goes well past them on both sides. A dark patch lies north of these, with some fainter nebulosity extending beyond it.

Just north of NGC 1977 we come to **NGC 1981,** another loose, sparse star cluster that's an easy find through binoculars or a small telescope. About a dozen 6th-through 10th-magnitude stars seem to trace the path of a bouncing ball about $\frac{1}{3}°$ wide from east to west. The three stars forming the eastern bounce are the brightest and are aligned almost north-south, forming the dim north end of Orion's Sword. Through a 4-inch telescope about a dozen fainter stars can be counted in the area.

The bottom end of the sword is marked by its brightest star, 3rd-magnitude **Iota (ι) Orionis,** $\frac{1}{2}°$ south of the Orion Nebula. Iota is an attractive triple star for small telescopes that appears white, blue, and orange-red to many observers. The brighter of Iota's two companions is magnitude 7.0 and lies 11″ to the southeast; it can be seen at 50×. Both it and the primary are hot, blue-white stars, but when they are seen close together like this the dimmer star usually appears much bluer by contrast. Iota's second companion is 10th magnitude; you'll probably need a 4-inch or larger scope to see it 50″ to the east-southeast. It is often perceived as some shade of red.

Just southwest of Iota is the wide double star **Σ747** (Struve 747), a pair of blue-white gems 36″ apart shining at magnitudes 4.8 and 5.7. Steadily held binoculars will resolve them.

Just west of this pair, by less than half their distance from Iota, is a dimmer but eye-catching double, **Σ745.** It consists of 8th- and 9th-magnitude stars 28″ apart. This is not a true binary but an optical pair of relatively nearby stars about 280 and 150 light-years distant.

This tour of Orion's Sword has spanned less than 2° of sky. If your telescope can yield a magnification as low as 25× in a good eyepiece, you can encompass the entire area in one field of view. It's one of the most magnificent showpieces of the night. Northern observers may shy away from the icy darks of January and February, but the winter sky beckons with some of the most beautiful vistas in the heavens. It is well worth bundling up to spend an evening among them.

The Hunter's Treasures

Object	Type	Mag.	Dist. (l-y)	RA	Dec.
M42	Diffuse nebula	—	1,500	5^h 35^m	−5.5°
Trapezium	Multiple star	5.1, 6.5, 6.7, 8.0	1,500	5^h 35.3^m	−5° 23′
M43	Diffuse nebula	—	1,500	5^h 35.3^m	−5° 16′
NGC 1977	Cluster + nebula	~4	1,500	5^h 35^m	−4.8°
NGC 1981	Open cluster	4.6	1,500	5^h 35^m	−4.4°
ι Ori	Triple star	2.8, 7.0, 9.7	1,500	5^h 35.4^m	−5° 55′
Σ747	Double star	4.8, 5.7	1,500	5^h 35.0^m	−6° 00′
Σ745	Optical double	8.4, 9.4	280, 150	5^h 34.8^m	−6° 00′

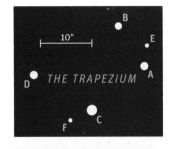

Below: This close-up of the inner Orion Nebula roughly matches the greenish hue seen by many visual observers. The stars of the Trapezium appear nearly lost in the nebula in this Lick Observatory photograph, but they are its brightest feature visually. Use the diagram at right to hunt for the fifth and sixth Trapezium stars, E and F, on a night of good seeing.

A Walk in Starry Mists

ORION, THE HUNTER, dominates the evening sky with unmatched brilliance. This constellation's premier treasure is the Great Orion Nebula, M42, an expansive swath of intricate nebulosity with a heart of sparkling gems. It is visible to the unaided eye as the middle "star" in Orion's Sword. See page 46 for details of its telescopic appearance and a list of several pretty multiple stars, including the Trapezium.

Although the Great Nebula is often the first object skygazers turn their scopes toward, Orion shelters many compelling deep-sky delights. Long-exposure photographs show that M42 is part of a large cloud of gas and dust encompassing much of the constellation. The Orion Cloud is approximately 1,500 light-years away and several hundred light-years across.

NGC 2024, one of the brightest patches in the Orion Cloud, lies right beside Alnitak, or Zeta (ζ) Orionis. My 4.1-inch refractor at 87× shows the Flame Nebula, about 25′ across with a network of dark lanes. Most obvious is a wide lane that plunges into the nebula from the northeast and exits to the south. Increasing the magnification to 153× shows that nearby Alnitak is a double star. It consists of a 3.7-magnitude companion 2.5″ south-southeast of the 1.9-magnitude primary.

Another patch of nebulosity 2.5° northeast of Alnitak was cataloged (like M42 itself) by 18th-century French comet hunter Charles Messier and thus bears the designation **M78.** Through my little scope at 68×, M78 is a

For a reason that should be obvious, NGC 2169 is also known as the 37 Cluster. Its colorful star patterns stand out well in this photograph by George R. Viscome of Lake Placid, New York. The field shown here is ¹/₃° tall.

ghost with glowing eyes. Two 10th-magnitude stars stare at us from a fan of nebulosity flowing away to the southeast. A wider and slightly brighter pair of stars lie 15′ north-northeast in the same field of view. The eastern one is enveloped in a small nebula, **NGC 2071.** Both are reflection nebulae, shining by the light of embedded stars. M78 contains a fairly rich star cluster. While it cannot be seen visually, it does show up beautifully in the Two Micron All Sky Survey (2MASS) infrared images, available at www.ipac.caltech.edu/2mass.

Barnard's Loop is the remarkably long arc of nebulosity that sweeps through Orion in a huge semicircle about 12° in diameter. It was formed by ionization and energetic wind- and supernova-driven outflows from the association of hot, massive stars contained in the Orion Cloud. One of the biggest challenges in trying to observe Barnard's Loop is having a large enough field of view. In fact, some observers have found it with the naked eye by gazing at the sky through a light-pollution filter. Others have taped filters over the objectives of their binoculars.

One of the brightest segments of Barnard's Loop is centered about 1.2° northeast of M78, and it responds particularly well to a hydrogen-beta filter. With my small refractor so equipped and an eyepiece giving a 3.6° field of view, I find this portion of Barnard's Loop surprisingly easy. It appears variable in width, averaging about 40′ across, and can be traced for several degrees. Some patches are brighter than others.

Enmeshed in Barnard's Loop is the open star cluster **NGC 2112,** 1.7° east of NGC 2071. At 127×, I see a very pretty scattering of many faint to extremely faint stars dusting a patch of sky about 10′ across. One brightish star lies just outside the northwest edge.

Although first noted by several astronomers in the early days of celestial photography, what we call Barnard's Loop has been seen visually (at least parts of it) with today's nebula filters.

Now let's move up to the distinctive open cluster **NGC 2169.** You'll find it 53′ west-southwest of Xi (ξ) Orionis, the star marking the Hunter's upraised, eastern hand. At 68× I see 16 stars of 7th magnitude and fainter in two groups, collectively 6′ across. The northwestern group contains six stars. The southeastern group contains 10, the brightest being a close double (Σ848 AB).

For observing comfort I usually use a star diagonal, which gives my refractor a mirror-image view. If I remove the star diagonal, it becomes obvious how this little cluster earned its nickname, the 37 Cluster. The larger bunch of stars clearly forms a pointy-lobed 3, and the smaller group makes a 7. In a telescope that gives an inverted image (south up), the 37 will be upside down. Alan Goldstein of Louisville, Kentucky, coined the name in 1981.

A much different cluster lies 1.5° south-southeast of Xi. With a small scope you'll need a high power to pick out the minute sparkles of light inhabiting **NGC 2194.** At 153× I can make out many faint to extremely faint stars over a misty, unresolved background about 7′ across. The brightest is a 10th-magnitude star at the southern edge of the group.

Another nice nebula, **NGC 2175,** sits 1.4° east-northeast of Chi^2 (χ^2) Orionis. In my little refractor at 47× I find the nebula obvious without a filter. However, a narrowband filter betters the view, and an oxygen III filter helps even more. An 8th-magnitude star is visible near the center, and a dozen faint stars are superposed. The nebula is slightly mottled and shows hints of dark lanes.

NGC 2175 is sometimes plotted as an open cluster in star atlases, while the designation NGC 2174 is given to the nebula. Neither characterization is correct. NGC 2175 was discovered sometime in the mid-1800s by German astronomer Carl Christian Bruhns and first reported by Arthur Auwers, who described it as an 8th-magnitude star within a large nebula. NGC 2174 is actually a bright knot of nebulosity in the northern edge of NGC 2175. It was discovered at Marseilles Observatory by Édouard Stephan, widely recognized for the group of galaxies that bears his name: Stephan's Quintet (in Pegasus). Folks with larger scopes might like to hunt for NGC 2174 and for the even brighter knot Sh2-252 E, located 11′ north-northwest and 3.3′ east-northeast, respectively, of the 8th-magnitude star.

The existence of an open cluster within the nebula seems debatable. It was Swedish astronomer Per Collinder who first noted a cluster here and mistakenly equated it with NGC 2175. The cluster's designation should properly be Collinder 84, but there doesn't appear to be an obvious concentration of stars within the nebula. Collinder 84 is supposed to consist of the clumps of stars loosely scattered across most of the nebula NGC 2175. Does it look like a cluster to you?

Orion's Less Well-Known Treats

Object	Type	Mag.	Size	Dist. (l-y)	RA	Dec.
NGC 2024	Emission nebula	10.5	30′ × 22′	1,500	5^h 41.7^m	−1° 48′
M78	Reflection nebula	8.0	8′	1,500	5^h 46.8^m	+0° 04′
NGC 2071	Reflection nebula	8.3	8′	1,500	5^h 47.1^m	+0° 18′
Barnard's Loop	Emission nebula	—	12°	1,500	5.3^h to 6^h	+3° to −9°
NGC 2112	Open cluster	9.1	11′	2,600	5^h 53.8^m	+0° 25′
NGC 2169	Open cluster	5.9	6′	3,400	6^h 08.4^m	+13° 58′
NGC 2194	Open cluster	8.5	10′	12,300	6^{h}h 13.8^m	+12° 48′
NGC 2175	Emission nebula	7.3	29′ × 27′	5,300	6^h 09.7^m	+20° 29′

Angular sizes are from catalogs or photographs. Most objects appear somewhat smaller when a telescope is used visually.

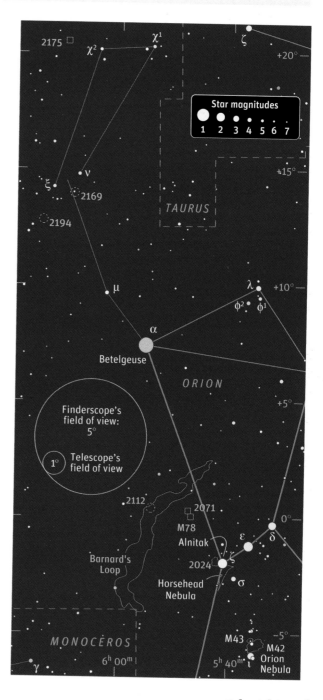

A Leap into Lepus

MANY ACQUAINTANCES know I have a deep fondness for rabbits. So what could be more natural than for me to combine this love with astronomy and visit the rabbit's closest celestial relative, Lepus, the Hare?

There are a number of stories associated with Lepus. My favorite involves the goddess of spring, whom the Anglo-Saxons called Eostre. In one version of the tale, the goddess created Lepus from a bird. In tribute to its former life, Eostre granted this hare the ability to run as fast as a bird can fly. She also allowed it to lay eggs

Tucked near the brilliant stars of Orion (top right) and Canis Major (left), the dim constellation of Lepus seldom attracts observers with small telescopes. Yet it is host to some rather unusual sights. The outline in this view locates the region shown on the facing chart.

on one day of the year, and so every Easter (which is derived from the name of this goddess) a bunny supposedly delivers our eggs.

The constellation Lepus has no stars brighter than 3rd magnitude. No doubt the hare is trying to escape attention as Orion, the Hunter, strides close by. Deep-sky enthusiasts, however, will find Lepus well worth notice.

First we'll visit **NGC 2017,** a star grouping that lies 1.6° east of the brightest star in Lepus, Alpha (α) Leporis. NGC 2017 is a near ghost town of a cluster. The mere nine members of this poorly populated collection are listed in the *Washington Double Star Catalog* as the components of a multiple-star system, but most of the stars are not related and just happen to lie along the same line of sight.

With my 4.1-inch (105-millimeter) refractor at 87×, I can see five or six stars arranged in two almost parallel lines running northwest to southeast. Despite its sparseness, NGC 2017 is a pretty group with the four brightest members displaying a nice variety of colors. In order of decreasing brightness, I see them as blue-white, orange, yellowish, and white.

Lepus has a number of pleasingly colored star pairs, among them the naked-eye star **Gamma (γ) Leporis.** It is a wide double easily split at 17×. I see a yellow-white 4th-magnitude star with a 6th-magnitude orange attendant to the north. This is one of the nearest star systems to Earth, 29 light-years, and its components are separated by at least 20 times the average distance of Pluto from our Sun.

Yellow Beta (β) Leporis, a 3rd-magnitude star, will serve as the jumping-off point for another pretty double, **h3759.** It lies 1.2° north-northwest of Beta and is the brightest star in the area. While it is a much closer pair than Gamma, h3759 is still well split at 17×. The 6th-magnitude primary has a 7th-magnitude secondary to the northwest. While both stars appear yellowish, the dimmer one seems to have a deeper hue. Parallax measurements with the Hipparcos spacecraft indicate that these are unrelated stars lying at quite different distances.

If you imagine a line from Alpha through Beta and extend it for that distance again, you will end up in the region of Lepus's sole Messier object, the globular cluster **M79.** It is framed to the north and south by two 9th-magnitude stars set 19′ apart.

M79 is a little fuzzy patch at low powers. At 153×, it looks fairly bright with a mottled core and a large halo. Many faint stars are just barely visible in the halo, and a 12th-magnitude star sits at its extreme northern edge. We see most of our galaxy's globular clusters when we look toward the galactic core (in the general direction of Sagittarius). However, M79 is beyond our Sun as viewed from the center of the galaxy, so we see it on the opposite side of our sky.

With a low-power eyepiece, you can fit M79 into the same field of view with a 5th-magnitude star 36′ to the west-southwest. This is the tight double star **h3752.** At 87×, it is a very close, bright pair of yellow and white suns. You may need powers of about 140× to comfortably split the duo.

In a more out-of-the-way corner of the constellation we find **R Leporis,** or Hind's Crimson Star, one of the reddest stars in the sky. By way of comparison, look at Orion just north of Lepus. Its two brightest stars are Rigel and Betelgeuse, well known for their lovely contrasting colors. Blue-white Rigel has a color index of –0.03, while that of decidedly orange Betelgeuse is +1.85. According to an article by Brian Skiff on carbon stars (*Sky & Telescope:* May 1998, page 90), Hind's Crimson Star has a startlingly red color index of +5.5!

Most such deeply red stars are variables. R Leporis is a Mira-type variable with a period of about 427 days. It is always within the reach of a small telescope, for its irregular magnitude extremes are 5.5 and 11.7. As its brightness changes, so does the intensity of its reddish color. Be sure to look at this star occasionally; is it brightening or dimming now?

To locate R Leporis, draw a line from Lambda (λ) through Kappa (κ) Leporis and continue for twice that distance again. This will bring you to a 4.8-magnitude star in neighboring Eridanus. From there, R Leporis is a straight drop 2.3° south. Can you see its color now, or will you have to wait for it to brighten again?

While we're in the northern reaches of Lepus, let's visit the unusual planetary nebula **IC 418.** Tiny but bright, IC 418 is found 2° east-northeast of

All of objects mentioned above can be seen (though some just barely) with a good set of 50-millimeter binoculars under a dark sky. But a small telescope will provide more satisfying views. This chart includes stars to magnitude 8.5 for help in star-hopping to their locations (for reference, α Leporis is magnitude 2.6).

Lambda. Through my 4.1-inch scope at 87×, the nebula looks small, round, and turquoise. The 10th-magnitude central star dominates the nebula. Observers with larger telescopes sometimes see the planetary as a pink, reddish, or pumpkin color. As far as I know, an 8-inch scope is the smallest instrument in which these unusual colors have been reported. Exercising my imagination, I've thought I could detect a pinkish tint in a 10-inch scope. Low powers seem to help.

As you can plainly see, Lepus, the Hare, has a fine variety of deep-sky wonders to offer. If you need a good observing project some clear February evening, hop outside (quick like a bunny) and leap into the treasures of Lepus.

Lepus' Sights for Small Scopes

Object	Type	Mag.	Size	Dist. (l-y)	RA	Dec.
NGC 2017	Asterism	—	10′	—	5^h 39.3^m	–17° 51′
γ Lep	Double star	3.6, 6.3	97″	29	5^h 44.5^m	–22° 27′
h3759	Optical double	5.9, 7.3	27″	140, 93	5^h 26.0^m	–19° 42′
M79	Globular cluster	7.7	9.6′	42,000	5^h 24.2^m	–24° 31′
h3752	Double star	5.4, 6.6	3.5″	290	5^h 21.8^m	–24° 46′
R Lep	Red variable star	5.5–11.7	—	800	4^h 59.6^m	–14° 48′
IC 418	Planetary nebula	9.3	12″	1,300	5^h 27.5^m	–12° 42′

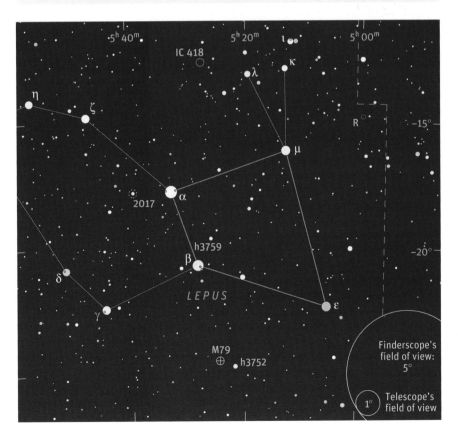

Hodierna's Auriga

FROM MIDNORTHERN LATITUDES, Auriga, the Chari-
oteer, can be found straight overhead on February
evenings. Its distinctive pentagon of bright stars with
brilliant Capella — also known as Alpha (α) Aurigae —
at one corner makes this an easy constellation to spot.
Auriga is extremely rich in deep-sky treasures, many
easily visible through small telescopes.

Notice that one of the stars in the pentagon seems to
be shared with the neighboring constellation Taurus,
the Bull. Long ago it was not uncommon for constellations to share a star, but when the International Astro-

The open clusters M38 (top) and NGC 1907 (bottom) lie buried in the heart of
Auriga. M38 has more than 300 stars, some arranged in chains radiating from
the center. NGC 1907 is a compact cluster of fewer than 100 stars about ½°
south-southwest of its companion.

nomical Union set the constellation boundaries in
1930, this star was placed within Taurus and is now officially Beta (β) Tauri.

This time we focus on **M36, M37,** and **M38** — three
open clusters discovered by the Sicilian Giovanni
Hodierna (1597–1660). A tract containing his observations was published in 1654 and rediscovered only recently. Hodierna classified these star clusters as *nebu-
losae* — patches of the sky that appear as small clouds
to the naked eye but show themselves to be made of
stars close together when seen through a telescope.

We'll start with **M38,** buried deep in the heart of the
pentagon. It is located 7.2° due north of Beta Tauri and
makes a nearly equilateral triangle with that star and
Iota (ι) Aurigae. M38 is visible through a finder as a
small hazy spot. With my 4.1-inch refractor at 68×, I
see about 60 stars of 8th magnitude and fainter in an
area around 20′ across, but the boundaries are ill defined. The cluster has a distinctive shape, with a lone
9.7-magnitude star centered in a 5′ hole nearly devoid
of stars. Many of the brighter stars radiate from this
central dark zone in four arms, making a cross with
the longer bar running west-southwest to east-northeast. M38's brightest member, an 8.4-magnitude yellowish star lying at the east-northeast edge of the cluster, marks the foot of the cross. Increasing the
magnification to 87× brings out some fainter stars,
swelling the star count to about 80.

At 87× my scope has a 53′ true field, which lets me
shoehorn M38 into the same view with neighboring
NGC 1907. This little star cluster lies south-southwest of
M38 and is about 6′ across. Through my refractor I see
a rich sprinkling of very faint stars over a hazy background, with one slightly brighter star near the center.
A matched pair of 10th-magnitude stars adorns the
south-southeast edge.

M36 lies a mere 2.3° southeast of M38. It makes a
very squat isosceles triangle with Beta Tauri and Theta
(θ) Aurigae and is visible through a finder. My refractor
at 87× shows 50 moderately bright to faint stars in a
loose group about 15′ across. Multiple curving arms
composed of the brighter stars radiate from the center.
You can see a pair of 9th-magnitude stars **(Σ737)** east-
southeast of center, and a wider pair of 10th-magni-
tude stars **(Sei 350)** south of center. The stars of M36
look as if they were scooped from a spot just south of
the cluster where we find a void of about the same
size. This is the dark nebula **Barnard 226,** which starts
about 10′ from the cluster's southern edge.

Our final open cluster, **M37,** is 3.7° east-southeast of
M36 and is the brightest of the bunch. Look for it approximately as far outside Auriga's pentagon as M36 is

inside it. M37 is the stunner of Hodierna's three Auriga clusters. My refractor at 87× reveals an extremely rich flurry of faint stars gathered in clumps scattered around a central, orangish magnitude-9.1 star. Dark lanes and patches abound, threaded among the teeming swarms of stars. This beautiful confusion of stars is one of my favorite deep-sky sights.

Giovanni Hodierna was one of the first astronomers in Sicily to grasp the importance of Galileo's new ideas. His instruments were simple Galilean refractors, the only known example having a magnification of 20× and a limiting magnitude of about 8. He systematically swept the sky, planning to publish a sky atlas with 100 maps on which he would include his nebulous objects. Hodierna never completed the work and only a few samples of his observations remain, which nevertheless appear to include 46 objects. Some are mere asterisms and some are not described well enough to identify, but of the "nebulae" he observed, at least 10 seem to be original discoveries. This is quite extraordinary when we consider that during the same youthful era of the telescope, the rest of the astronomical community discovered just one such object, the Orion Nebula.

Hodierna believed that "all the admirable objects that can be seen in the sky" could be resolved into stars if it were not for the limitations of his telescope. He speculated that the apparent brightness of these stars could depend not only on their intrinsic brightness but also on their distances. He entertained the idea that the distribution of stars might seem disordered to our view because they were ordered about some other spot in the universe. As a religious man in an era when putting the Sun at the center of the universe was still dangerous, Hodierna was quite daring to discuss, even hypothetically, a center much farther away.

Due to Hodierna's relative isolation and a general lack of interest in stellar astronomy at the time, his discoveries and ideas have remained nearly unknown. Let's remember them as we gaze at the clusters of Auriga and appreciate the fact that he would have welcomed the humblest of our telescopes as marvelous instruments.

Detail at right: Small-scope users looking for an observing challenge can turn their gaze toward the center of M36, where the two close double stars Σ737 and Sei 350 reside. On the finder chart at far right (and in all photographs) north is up and east is left. The circles show fields of view for a typical finderscope (5°) or a small telescope with a low-power eyepiece (1°). To find north through your eyepiece, nudge your telescope toward Polaris; new sky enters the view from the north edge. (If you're using a right-angle star diagonal it probably gives a mirror image. Take it out to see an image matching the map.)

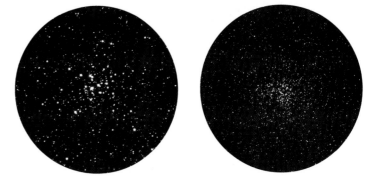

Left: Like its companions in Auriga, M36 looks somewhat like a hazy star against the backdrop of the Milky Way. Increased magnification opens up its crowded core. *Right:* The loveliest and most richly populated cluster in Auriga is M37. Look for an orange star at its heart, surrounded by clumps of stars and dark lanes.

Treasures of Auriga

Object	Type	Mag.	Dist. (l-y)	RA	Dec.
M38	Open cluster	6.4	4,300	05ʰ 28.7ᵐ	+35° 50′
NGC 1907	Open cluster	8.2	4,500	05ʰ 28.0ᵐ	+35° 19′
M36	Open cluster	6.0	4,000	05ʰ 36.0ᵐ	+34° 08′
Σ737	Double star	9.1, 9.2	4,000	05ʰ 36.4ᵐ	+34° 08′
Sei 350	Double star	10.3, 10.3	4,000	05ʰ 36.2ᵐ	+34° 07′
B226	Dark nebula	—	—	05ʰ 37.0ᵐ	+33° 45′
M37	Open cluster	5.6	4,400	05ʰ 52.5ᵐ	+34° 33′

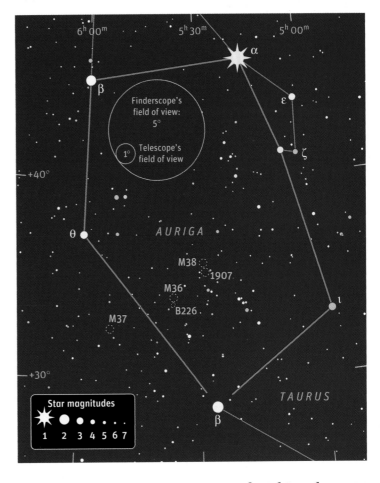

Double Your Fun

YOU'LL FIND GEMINI, the Twins, high in the south on the all-sky chart on page 22. Its brightest stars, nearly at the zenith, bear the names of the mythical brothers. The star Castor marks the head of one twin, while Pollux marks the head of the other. Gemini harbors one of the finest open clusters, most spectacular planetary nebulae, and most fascinating multiple-star systems in the northern sky.

We'll begin with beautiful **M35,** which Massachusetts amateur Lew Gramer calls the Shoe-Buckle Cluster. It glitters on Castor's northern "shoe," between and above Eta (η) and 1 Geminorum. M35 is faintly visible to the unaided eye under dark, transparent skies and is clearly evident through a small finder as a patch of mist. I find it quite striking with 7 × 50 binoculars and partially resolved into moderately bright stars. In his 1888 classic, *Astronomy with an Opera-Glass,* Garrett Serviss described this cluster as "a piece of frosted silver over which a twinkling light is playing."

My 4.1-inch (105-millimeter) refractor at 47× shows this to be an outstanding cluster with more than 100 stars in an area the size of the full Moon. There is a void near

Striking for its colors, the cluster NGC 2266 is seen here as no small scope can ever show it. The ¼° field was recorded with the Calar Alto 1.23-meter reflector in Spain.

the center populated by only a few faint points of light. A pretty arc of stars starts at the north edge of M35 and curves toward the west as it reaches the void. My impression through my 6-inch reflector at 39× is a bit different. Here I note a 70-star core in a vaguely rectangular group about 20′ by 25′ running northwest to southeast. The bright star in the northeast side has a distinctly golden hue; it is the primary component of the double star OΣ **134,** with a bluish companion to its south.

Recent studies give M35 an age of about 150 million years. On the cosmic or even geologic time scale, this is quite young. While gazing at M35, you might be stepping on stones older than that.

The little fuzzy spot next to M35's fringe is the open cluster **NGC 2158.** Zooming in at 127×, my little refractor shows many extremely faint points of light over a hazy, unresolved background 4′ across. The brightest star lies at its southeast edge. This would be a very impressive-looking cluster if not for its immense distance of 12,000 light-years — more than four times as far off as M35. NGC 2158 is a very old cluster with an age of around 2 billion years; the most prominent stars sprinkled across its face are red giants.

NGC 2266 is a comparable group sitting 1.8° north of Epsilon (ε) Geminorum, the star that marks Castor's knee. At 87×, I see a dozen faint to very faint stars against a background haze from which elusive pinpoints of light wink

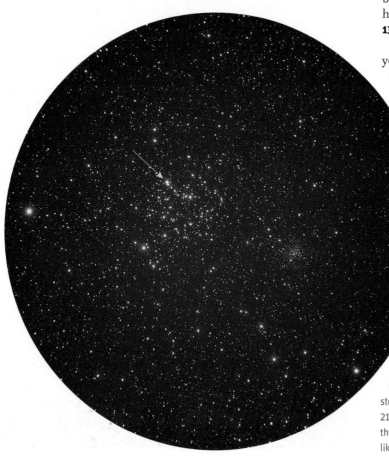

Gemini's best-known star cluster, M35, is the broad sprinkling of stars left of center. Its much more remote "companion" cluster, NGC 2158, often eludes observers in light-polluted skies. An arrow marks the double star OΣ 134 in M35. The field shown is nearly 3° across, like that presented by a typical 20× spotting scope; north is up.

in and out of view. This pretty cluster is 6' across and somewhat triangular, with the brightest star at the southwestern point. NGC 2266 contains many red-giant stars, and in color images it is one of the most beautiful star clusters I've ever seen.

Now let's hop over to Castor's twin, Pollux, where we'll find the double star **Delta (δ) Geminorum.** The planet Pluto was quite close to this star when it was discovered on photographs taken in January 1930. At 87× with my 4.1-inch scope, Delta is a very tight pair. The yellow-white primary has an 8th-magnitude attendant to the southwest. Although this companion is a red-dwarf star, some observers describe it as red-purple or even blue.

The intricate planetary nebula **NGC 2392,** or Caldwell 39, dwells 2.4° east-southeast of Delta. An 8th-magnitude star just north of the planetary makes the pair resemble a double star when viewed at low power. My little scope at 153× reveals a small, roundish, slightly mottled, blue-gray glow surrounding a bright central star. Some skygazers notice a blinking effect when looking at this nebula. Using averted vision makes the nebula more apparent, while staring straight at it makes it blink off.

NGC 2392 is often called the Eskimo Nebula because some ground-based photos make it look like a face encircled by a furry hood. The Eskimo's fur parka may be detected as a decrease in brightness through small scopes, but you'll probably need at least an 8-inch to pick up any of the dark ring separating it from the Eskimo's face. A 6-inch scope begins to show some of the dark features in the Eskimo's face, notably a dark patch west of center that outlines part of the nose. When it comes to viewing the subtle details of the Eskimo Nebula, high power coupled with a light-pollution filter of the oxygen III or narrowband type can be a great help.

In January 2000, the newly refurbished Hubble Space Telescope turned its gaze toward the Eskimo Nebula, revealing a fascinating wealth of detail. The structure of the fur hat is thought to be caused by a slow equatorial wind from the dying central star, while the filamentary face is molded by faster winds blowing from its poles (http://hubblesite.org/gallery).

Our final target is **Castor,** a system of six suns orbiting each other in an intricate ballet. Each of its three visible components hides a companion too close to be resolved through a telescope — three sets of secret twins disclosed only by their telltale spectra. Through my small refractor at 87×, 2nd-magnitude Castor A closely guards 3rd-magnitude Castor B to the east-northeast, and both appear white. Much dimmer Castor C lies a generous 71″ to the south-southeast. It consists of a pair of nearly matched red-dwarf stars undergoing mutual eclipses, so that we witness two during each orbital period of 19.5 hours. In each case the stars' combined light is halved, dropping about 0.7 magnitude. As a variable star, Castor C is designated **YY Geminorum.**

In the Constellation of the Twins

Object	Type	Mag.	Size/Sep.	Dist. (l-y)	RA	Dec.
M35	Open cluster	5.1	28′	2,700	6^h 09.0^m	+24° 21′
OΣ 134	Double star	7.6, 9.1	31″	2,700	6^h 09.3^m	+24° 26′
NGC 2158	Open cluster	8.6	5′	12,000	6^h 07.4^m	+24° 06′
NGC 2266	Open cluster	9.5	6′	11,000	6^h 43.3^m	+26° 58′
δ Gem	Double star	3.6, 8.2	5.5″	59	7^h 20.1^m	+21° 59′
NGC 2392	Planetary nebula	9.2	47″ × 43″	3,800	7^h 29.2^m	+20° 55′
Castor	Triple star	1.9, 3.0, 8.9	4.1″, 71″	52	7^h 34.6^m	+31° 53′
YY Gem	Variable star	8.9–9.6	—	52	7^h 34.6^m	+31° 52′

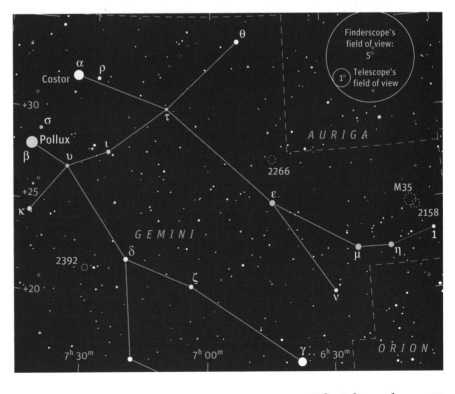

NGC 2392 (also known as the Eskimo Nebula) is quite tiny; the 8th-magnitude star in this view is just 99′ north of the nebula's central star.

Winter Clusters Galore

MUCH OF THE MILKY WAY is spangled with bright naked-eye stars, but not all of it. Examine the March all-sky chart on page 22. Southward from the feet of Gemini through Monoceros and northern Puppis, the winter Milky Way displays no star as bright as 3rd magnitude. Distracting the eye to the west of this region is bright Canis Major with Sirius. To the east is Canis Minor with bright Procyon. But don't be fooled; the "empty" sky between them is rich territory for deep-sky hunters with small telescopes.

M50 is a cluster you probably don't know about, but it's easy to locate. Draw a line from Sirius in Canis Major through the Big Dog's nose, Theta (θ) Canis Majoris, and continue onward for nearly the same distance (about 4°). M50 is a faint smudge in a good finderscope, and a 2.4-inch (60-millimeter) telescope shows a group of about 20 stars. Through a 3.6-inch, the cluster's ¼° core reminds me of a housefly facing northeast. One wing goes off toward the south and the other to the west. Stars on the northeastern side of the cluster form the head and mandibles. M50's brightest star lies in the southern wing; it looks reddish orange through telescopes larger than 4 inches.

M50 reminds me of a connect-the-dots housefly. Such eyepiece impressions rarely match how star clusters look in photographs, because bright and faint stars appear much more alike on photographs than they really are. The eye gives a truer view. Martin Germano used an 8-inch f/5 reflector for this 30-minute exposure on hypersensitized Kodak Technical Pan 2415 film.

Below: The field of M47 and M46 in northern Puppis. Note the tiny red ring of the planetary nebula NGC 2438 in M46. Chris Cook digitally enhanced and composited two 25-minute exposures made with an 85-mm f/5.6 refractor to create this image.

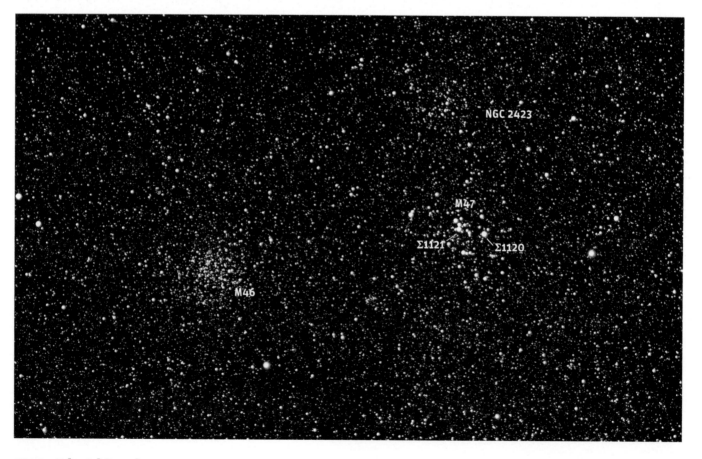

Draw a line from Mirzam — Beta (β) Canis Majoris — through Sirius and continue east for a little more than twice that distance, and you'll land just north of **M47**. Dark skies reveal a very faint fuzzy patch to the unaided eye. A good finder will pick it up under suburban skies. M47 is a beautiful cluster through any telescope. In 14 × 70 binoculars, 25 bright to faint stars can be seen in a very loose, irregular group about ¹/₂° across. In my 4.1-inch M47 shows about 50 mixed bright and faint stars weakly concentrated toward the center.

An arc of three fairly bright stars lies just north of center. The southernmost of the three is the double star **Σ1121** (Struve 1121), a nearly matched pair of 7th-magnitude stars separated by 7.4″. They can be cleanly split at 75×. M47's brightest star, near the western edge, is also double; **Σ1120** has 5.7- and 9.6-magnitude suns separated by 19″. The brightest star north of Struve 1120 is orange.

The open cluster **M46** lies just 1.3° east-southeast of M47 and offers a study in contrasts with it. M46 appears much smoother and finer grained than its coarse, bright-starred neighbor. Look for a glowing swarm of hundreds of faint stars, many just on the edge of visibility through a small scope. In my 4.1-inch M46 appears round and extremely rich, with many faint to very faint stars densely packed over an unresolved background.

As an added bonus, the little planetary nebula **NGC 2438** is clearly visible as a small, round glow in M46's northeastern reaches. You'll need a magnification of 80× or more to recognize the nebula. Early research indicated that NGC 2438 was a foreground object, but its exact distance remains uncertain.

There's a third cluster here too. Just 40′ north of M47 a 2.4-inch telescope will pick out a small, faint patch of light — **NGC 2423.** Much subtler than its splashy neighbors, NGC 2423 looks about half their size. Through my 4.1-inch scope it shows about 20 dim stars against an unresolved haze.

Together M47, M46, and NGC 2423 make up one of the prettiest fields in the sky for small rich-field scopes. At 25× or less all three will fit in the same field of view. Because they appear strikingly different, the total effect is breathtaking.

It is even more fascinating if we keep in mind the trio's relative placement in space and time. M47 is the closest and youngest at about 1,600 light-years and age 130 million years, while NGC 2423 is approximately seven times as old and 1¹/₂ times more distant. M46 is about the same age as M47 but nearly three times farther away, at around 4,500 light-years.

The cluster **M48** is 12° farther northeast. It makes a not-quite-equilateral triangle with the 4th-magnitude stars C Hydrae (a wide triple in binoculars or a finderscope) and Zeta (ζ) Monocerotis. A 2.4-inch shows a big, poorly defined group 40′ wide made of about 30 stars 8th magnitude and fainter. The brighter stars run north-south across the center. A 4-inch scope shows 50 stars, many arranged in pairs and chains.

Our celestial Dogs have herded a flock of starry treasures to view on March nights. In my next essay I'll zoom in on Sirius — the brightest star in the Larger Dog — and its Pup.

Some Winter Treasures

Object	Type	Mag.	Dist. (l-y)	RA	Dec.
M50	Open cluster	5.9	3,000	7ʰ 03.2ᵐ	–8° 20′
M47	Open cluster	4.4	1,600	7ʰ 36.6ᵐ	–14° 30′
Σ1121	Double star	7.0, 7.3	1,400	7ʰ 36.6ᵐ	–14° 29′
Σ1120	Double star	5.7, 9.6	1,400	7ʰ 36.1ᵐ	–14° 30′
M46	Open cluster	6.1	4,500	7ʰ 41.8ᵐ	–14° 49′
NGC 2438	Planetary nebula	11	2,900?	7ʰ 41.8ᵐ	–14° 44′
NGC 2423	Open cluster	6.7	2,400	7ʰ 37.1ᵐ	–13° 52′
M48	Open cluster	5.8	2,000	8ʰ 13.8ᵐ	–5° 48′

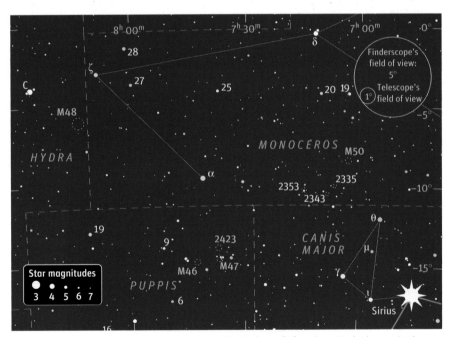

The circles show the size of a typical finderscope's field of view (5°) and a typical telescope's view with a low-power eyepiece (1°). This indicates how much of the map you'll see in your eyepiece. Next, figure out directions. North is up and east is left on the map. To find which way is north in your view, nudge your telescope slightly toward Polaris; new sky enters the view from the north edge. Turn the map around to match. (If you're using a right-angle star diagonal at the eyepiece, it probably gives a mirror image. Take out the diagonal to see a correct image that will match the map.)

Dog Overboard!

YOU CAN SEE the great ship Argo Navis sinking into the southern horizon on the March all-sky chart on page 22. This celestial ship was one of the 48 constellations in Ptolemy's Almagest, representing the famous vessel of the Argonauts. But modern astronomers have dismantled Argo Navis, or simply Argo, into smaller constellations outlining its keel, sails, and stern.

For those of us living at midnorthern latitudes it looks as if a disaster is occurring. Most of the great ship is out of sight below the southern horizon, with only the stern, Puppis, sticking up. Canis Major, the Larger Dog, appears to have jumped off the back of the doomed craft, and we see him in mid-leap to the northwest of Puppis.

In this area of the sky there are dozens of deep-sky delights within the grasp of a small telescope. Let's visit a few that are conveniently placed near naked-eye stars, starting our voyage in Canis Major at **Sirius**, the brightest star in the night sky.

Sirius is a double star, its white-dwarf companion having a diameter comparable to Earth's. Since Sirius is often called the Dog Star, its tiny companion has been nicknamed the Pup. The pair is not generally considered

game for small scopes. Although the Pup is fairly bright at magnitude 8.5, it is usually lost in the intense glare of brilliant Sirius. The apparent separation of the two stars is increasing, however, and in 2001 I was finally able to catch a glimpse of the Pup with my 4.1-inch refractor on one night of exceptionally fine seeing. By 2010 the Pup will be 8″ east of Sirius, and in 2022 it will reach its maximum separation of 11.3″ toward the east-northeast. What's the first year when you'll spot the Pup? Sighting this white-dwarf star definitely belongs on your list of greatest challenges.

Split or not, Sirius is the beacon that will guide us to the large open cluster **M41**. It lies 4° south of Sirius and is visible to the unaided eye from a dark observing site. Binoculars or finderscopes show M41 as a good-size hazy patch and may unveil some of its stars. Massachusetts amateur Lew Gramer calls this cluster the Dog's Spot. My 4.1-inch refractor at 17× shows 50 fairly bright to faint stars within 40′. The two brightest dwell in the heart of the cluster and form a wide, nearly matched pair with the western one displaying a golden hue. Some skygazers describe this star as orange or even reddish. What color do you see? Increasing the power to 47× helps bring out the fainter members, swelling the crowd to 80. Many of the stars seem to be arranged in chains radiating from the center, and a 6th-magnitude star (12 Canis Majoris) lies just off the southeast edge.

Our next guidepost is the 3.9-magnitude star Omicron¹ (o¹) Canis Majoris, which dwells within the boundaries of **Collinder 121**. While a little bigger and brighter than M41, Cr 121 looks far less like a cluster. Its stars are loosely scattered and show no central concentration. In my 4.1-inch scope at 17×, Omicron¹ looks deep yellow. A few dozen stars of 6th magnitude and fainter are gathered around it in a group with ill-defined borders. A bright star occupies the southern reaches of the cluster; a slightly dimmer one sits in the north. A kite-shaped group of 7th- and 8th-magnitude stars is conspicuous to the west of Omicron¹ along with a triangle of fainter stars to its east-southeast.

To locate our next cluster, look for the stars Delta (δ) and Tau (τ) Canis Majoris shining at magnitude 1.8 and 4.4, respectively. **NGC 2354** lies about halfway between them. At 47×, my little refractor shows 30 stars, faint to very faint, in a patch about 20′ across. This is not an impressive group, but its knots and curving lines of stars invite dot-to-dot games. To me it looks like a slug executing a sharp turn, while others claim to see a pattern like the constellation Scorpius.

Dazzling Sirius (upper right) and Canis Major seem to leap from Puppis, the stern of the now-defunct celestial ship Argo Navis. *Inset:* Because Sirius outshines it by 10,000 times, the Dog Star's white-dwarf companion is always a challenge for telescopes. When R. B. Minton took this photograph in 1972 with the 61-inch Catalina reflector, the two stars were separated by 11.3″. After closing to 2.5″ in 1993, they have begun to widen again in their 50-year cycle.

A pretty chest of starry gems surrounds the star Tau CMa itself. Tau rests like a bright sapphire amid a tiny bed of lesser jewels. The collection is generally known as the Tau Canis Majoris Cluster and bears the designation **NGC 2362.** Use medium to high powers to separate blue-white Tau from its host of close companions. My 4.1-inch scope shows about 25 stars crowded into a mere 6′. With an age of about 5 million years, NGC 2362 (also called Caldwell 64) is one of the youngest star clusters known. Tau, a blue supergiant of spectral class O9, is thought to be a true member of the group. At the cluster's distance of 4,800 light-years, Tau must shine with the light of 50,000 Suns.

Finally, we leave the Big Dog behind and jump onto the sinking stern of Argo. The yellow star Xi (ξ) Puppis, at magnitude 3.3, anchors the northeastern end of a small curve of three stars. The other two shine at magnitude 5.3 and 7.9 and are dressed in a paler yellow light.

Our last cluster, **M93,** lies 1.5° northwest of Xi. M93 is a misty blur when viewed in binoculars or a finder. The cluster looks very nice through my little refractor at 47×, appearing rich in 8th-magnitude and fainter stars. At 87× I count 33 stars in a 7′-by-10′ core, which is surrounded by a sparser, ill-defined halo containing about 50 stars of 10th magnitude and dimmer. The cluster is very patchy, with many irregular bunches of stars. The core looks roughly like a notched arrowhead, its two brightest stars being slightly orange. Other observers have imagined these stars arranged in the shape of a starfish or a butterfly. What does M93's pattern remind you of?

The open clusters described in this essay span a wide range of size,

richness, and brightness. We can delight in the diversity that makes this class of objects so much fun to explore. But wait, there's more. Turn the page to discover four additional open clusters located just east of Sirius. Be sure to enjoy the sights of Canis Major and Puppis before they completely submerge below the horizon in late spring.

Stellar Sights in and near Canis Major

Object	Type	Mag.	Size/Sep.	Dist. (l-y)	RA	Dec.
Sirius	Double star	−1.46, 8.5	7.2″	8.6	6ʰ 45.2ᵐ	−16° 43′
M41	Open cluster	4.5	38′	2,300	6ʰ 46.0ᵐ	−20° 44′
Cr 121	Open cluster	2.6	50′	1,500	6ʰ 54.1ᵐ	−24° 15′
NGC 2354	Open cluster	6.5	19′	13,000	7ʰ 14.3ᵐ	−25° 44′
NGC 2362	Open cluster	4.1	7′	4,800	7ʰ 18.8ᵐ	−24° 57′
M93	Open cluster	6.2	21′	3,400	7ʰ 44.6ᵐ	−23° 52′

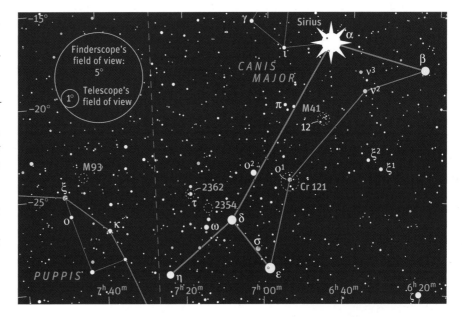

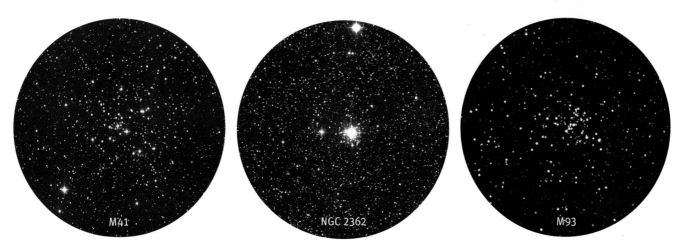

M41

NGC 2362

M93

Perambulations in Puppis

AT THIS TIME OF YEAR, wandering through Puppis is one of my favorite pastimes. Its exceptionally rich star fields provide fertile hunting grounds for any deep-sky observer, with dozens of objects visible through a small scope. One of the constellation's most magnificent areas is visible to the unaided eye as a hazy patch in a dark velvet sky. It contains most of the objects we'll take a look at in this essay.

Let's begin with **M46,** a singularly stunning star cluster. Even in 14 × 70 binoculars it appears extremely rich in minute stars. But to be sure you're looking in the right spot, first center your telescope on the 4th-magnitude star Alpha (α) Monocerotis and then slew 5° due south. My 4.1-inch (105-millimeter) refractor at 17× shows M46 as a round, densely packed swarm of faint stars over a hazy background. The brightest star is magnitude 8.7, and it's found in the western side; most of the rest are magnitude 11 or 12. Higher magnifications transform the misty backdrop into additional stars, but a small void rests at the cluster's heart.

While viewing this cluster at 87× I can see **NGC 2438,** a planetary nebula that seems to be embedded in its northern fringes. It appears round and fairly bright, and there's an 11th-magnitude star at its southeastern edge. NGC 2438 is 1' across, a little smaller than the famous Ring Nebula in Lyra. If you have trouble spotting it, a narrowband or oxygen III light-pollution filter will dim the stars and help the nebula stand out.

M47 is another fascinating star cluster. When he cataloged this object in 1771, Charles Messier described it as a cluster of stars not far from M46 and containing brighter stars. But he placed it at right ascension 7^h 54.8^m, declination –15° 25' (equinox 2000.0), which

Open star clusters M47 (the bright stellar group just below center) and NGC 2423 (the small, faint gathering of stars at the top) can be studied together with a low-power telescope. The field of view in this photograph is 1° tall; north is up. M47 is a pretty mix of bright and faint stars, but most of the members of NGC 2423 are 11th and 12th magnitude.

marks a conspicuously blank spot in the sky. Nonetheless, Johann L. E. Dreyer assigned the designation NGC 2487 to this location when he compiled his 1888 *New General Catalogue of Nebulae and Clusters of Stars.*

In 1934 German astronomer Oswald Thomas pointed out that another object in Dreyer's compilation, NGC 2422, situated 1.3° west-northwest of M46, must be what Messier really saw. In 1959 T. F. Morris of Montreal, Canada, offered a clever explanation for the confusion. Messier wrote that he had measured M47's location relative to the star 2 Navis (now known as 2 Puppis). The position he gave was 9^m east and 44' south of this star. If we say instead that M47 is 9^m *west* and 44' *north* of 2 Puppis, we end up fairly near the position of NGC 2422. So Messier may have applied each offset backward.

Whatever the story behind Messier's discovery, Sicilian astronomer Gioanbatista Odierna (also spelled Hodierna; see page 52) scooped him by more than a century. Odierna's little-known deep-sky catalog, published in 1654, both describes and maps the cluster we now call M47.

M47 is a beautiful sight in almost any instrument. Through 14 × 70 binoculars, 20 to 25 stars gather into a very loose, irregular cluster ½° across. I count 48 mixed bright and faint stars in my little refractor at 17×. Most of the dominant gems sparkle with fierce, bluish white star fire, but a few orange jewels can be picked out among them.

M47 contains several multiple stars. Perhaps the prettiest for a small scope is Σ1121, near the cluster's center. It is easy to single out as the southernmost in an arc of three bright stars. Σ1121 consists of a nearly matched pair of 7th-magnitude blue-white suns nicely split at 68× (see page 57).

Can you spot NGC 2438 as a tiny "soap bubble" in this 1°-wide photograph of the open cluster M46? Can you see it in your scope? Despite its apparent location within M46, the planetary is probably a foreground object that happens to lie along the same line of sight.

Another cluster can be found near M47; just follow a little chain of roughly 9th-magnitude stars northward. At 87× **NGC 2423** shows about 30 faint stars loosely scattered across 15'. The center of the group is fixed by one of its brightest members, the double star h3983 (a discovery of John Herschel). The 9.1-magnitude primary hosts a 9.7-magnitude secondary 8" west-northwest. The primary star is itself a double, but the components are too close for a small telescope to split. NGC 2423's lucida (an old term for "brightest member") is an 8.6-magnitude star at its south-southwest border, but most of the visible stars shine at magnitude 11 or 12.

A small scope with a short focal length can showcase all three clusters in the same field of view at 25× or less. Some wide-angle eyepieces will allow you to push the magnification as high as 45× and still fit them all in. The combination of these three disparate "star cities" makes for a captivating view.

Ranging a bit farther afield, we find the pretty little cluster **Melotte 71** 1.8° north of NGC 2423. With my 4.1-inch scope at 87×, I get the impression of a diamond crushed into fine powder with surviving chips casting glints of light. Near the southwestern edge gleam the two brightest stars, of which the eastern one is a close, matched double. Dimmer specks glitter against a frosty backdrop 9' across.

By returning to M46 and dropping 3.4° due south, we can turn up another planetary nebula. Look for it just west of an 8th-magnitude orange star. At low powers **NGC 2440** looks like a slightly fainter star, but 87× reveals a northeast-to-southwest oval that is distinctly robin's-egg blue. The planetary is bright and takes magnification well. At 153× I can see a bright spot in the middle, but this is not the central star. With his 4-inch refractor at about 400×, California amateur Ron Bhanukitsiri has been able to split this brightening into two distinct knots — a sight usually reserved for larger telescopes. Although too faint to be seen in a small telescope, the nebula's central star has one of the hottest confirmed surface temperatures. It blazes at more than 200,000°C, which is 30 times the surface temperature of our Sun.

As the lustrous Milky Way plunges through Puppis, it leaves many such wonders in its wake. Be sure to set aside the enchanting hours of a starry, moonless eve to enjoy them.

Sights in Northwestern Puppis

Object	Type	Mag.	Size/Sep.	Dist. (l-y)	RA	Dec.
M46	Open cluster	6.1	27'	4,500	7ʰ 41.8ᵐ	–14° 49'
NGC 2438	Planetary nebula	11.0	64"	2,900	7ʰ 41.8ᵐ	–14° 44'
M47	Open cluster	4.4	29'	1,600	7ʰ 36.6ᵐ	–14° 29'
NGC 2423	Open cluster	6.7	19'	2,500	7ʰ 37.1ᵐ	–13° 52'
Mel 71	Open cluster	7.1	9'	10,300	7ʰ 37.5ᵐ	–12° 03'
NGC 2440	Planetary nebula	9.4	20" × 15"	3,600	7ʰ 41.9ᵐ	–18° 13'

Angular sizes are from catalogs or photographs; most objects appear somewhat smaller when a telescope is used visually.

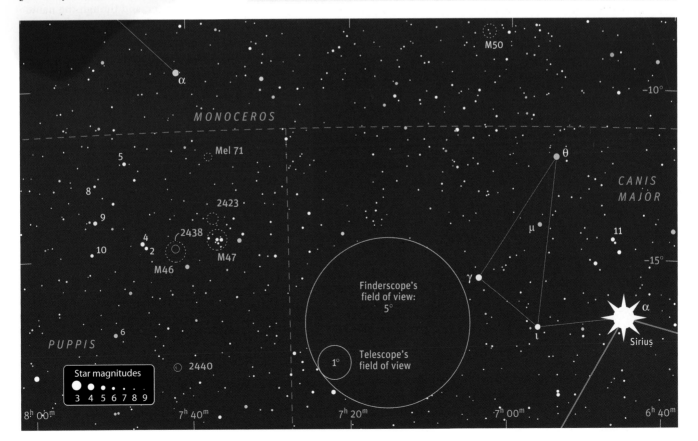

What a Crab!

THE ZODIACAL CONSTELLATION Cancer, the Crab, is now well placed for observing with its upside-down **Y** sandwiched between the brighter stars of Leo, the Lion, and Gemini, the Twins. In Greco-Roman mythology, the goddess Hera sent the Crab to attack the hero Hercules as he battled the multiheaded Hydra. Hercules promptly crushed this minor distraction underfoot, but Cancer gained a place in the sky for its trouble.

For the ancient Greeks the Sun appeared in Cancer on the summer solstice (June 21st). This placed the Sun directly overhead at noon at 23.5° north latitude, a line since known as the Tropic of Cancer despite the fact that the Sun now rests near the Taurus-Gemini border on that date.

Although this inconspicuous constellation contains no star brighter than 4th magnitude, it holds many interesting sights for the deep-sky observer. Each of the stars in Cancer's **Y** shape is a double, but none is more beautiful than **Iota¹ (ι¹) Cancri** at the figure's northern end. Through my 4.1-inch (105-millimeter) refractor at 17× I see the 4.0-magnitude primary as yellow, while the 6.6-magnitude secondary 30″ to the northwest has a nicely contrasting bluish cast.

Moving southward we come to a pair of double stars, **Gamma (γ) Cancri** (Asellus Borealis) and **Delta (δ) Cancri** (Asellus Australis), whose common names mean Northern Ass and Southern Ass. The gods Hephaestus and Dionysus rode these mythological donkeys into a battle between the Olympians and the Giants. When the Giants heard the braying of the approaching animals, they feared some terrible beast was being brought to destroy them and fled. The Aselli were placed in the sky to feed at a heavenly manger.

If you look just to the right of a line between Gamma and Delta, you will find that manger (called Praesepe in Latin). It is the star cluster **M44**, faintly visible as a hazy patch spanning about one-third the distance between those two stars. M44 is also known as the Beehive Cluster, and though the name is thought to be of more recent origin, Praesepe can also mean hive, possibly inspiring this alternate designation.

When the sky is moisture laden, the cluster fades from view. Under exceptional conditions, people have been able to resolve stars in the Beehive with the unaided eye. The explorer Thomas James recorded his observations in January 1632. During a search for a Northwest Passage while wintering on Charlton Island in what was later, in his honor, named James Bay he wrote:

> The thirtieth and one and thirtieth, there appeared, in the beginning of the night, more Starres in the firmament than euer I had before seene by two thirds. I could see the Cloud in Cancer full of small Starres, and all the via lactea nothing but small Starres; and amongst the Plyades, a great many small Starres. About tenne a Clocke, the Moone did rise, and then a quarter of them was not to be seene.

We may never behold the Beehive, the Milky Way, and the Pleiades so resplendent across a pristine sky, but catching the glint of stars within the Beehive has also been accomplished by modern observers at high altitudes away from light pollution.

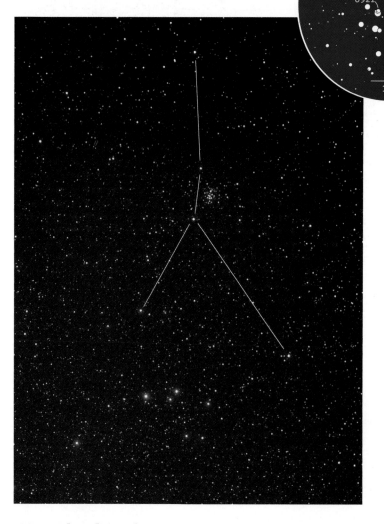

Cancer is one of the first starry harbingers of spring. It scuttles across the sky chasing after Gemini and just ahead of Leo.

M44, also called the Beehive Cluster, was known variously to the ancients as the Little Cloud, the Little Mist, Nubilum, and Praesepe (the Manger). Its 200+ stars lie 520 light-years away.

veals a rich group of 50 faint to very faint stars in about 25′. A 7.8-magnitude yellow-orange star lies just within the northeastern edge, while the rest range from 9th through 12th magnitude.

M67 has an irregular outline that has stirred the imagination of many observers. Admiral Smyth described it as "gathered somewhat in the form of a Phrygian cap; followed by a crescent of stragglers," an image unlikely to leap to mind. Camille Flammarion likened it to a sheaf of corn, a more manageable semblance. Stephen James O'Meara sees either a rearing king cobra that has swallowed its prey, or a swarm of insects attracted to the golden light of the brightest star. Unleash your imagination while gazing at the suns of M67, and see what impression they leave in your mind.

Those fettered with more typical observing conditions will need some optical aid to see the stars of M44. Even a small finder can reveal 20 stars. Visible in 14 × 70 binoculars are about 20 bright stars and 30 dimmer ones in a loose gathering nearly 1.5° across. With my 4.1-inch refractor at 47×, I can count 80 stars on an average night from my semirural home in upstate New York. On nights of very good transparency, I have picked out 150.

M44's stars are arranged in many distinctive triangles and bunches, with a **V**-shaped asterism reminiscent of the Hyades just south of center. The low-power multiple **ADS 6915** marks the point of the **V**. The easiest members to spot are a set of 6.7, 7.3, and 7.5-magnitude suns that range in color from yellow-orange to yellow-white. The quadruple star **ADS 6921** lies in the northern arm of the **V**. All four components can be seen at 47×. They shine at magnitudes 6.4, 7.6, 9.2, and 10.4. The two brighter stars appear orange and white, while the colors of the dimmer components are difficult to discern through a small telescope. The orange star makes a nearly equilateral triangle with two of similar hue and brightness 21′ north and 24′ east-northeast. A larger scope will pick out the colors more easily but will usually not allow a wide enough field to encompass the entire cluster.

Cancer contains a second Messier cluster, **M67**, often passed by in favor of its brighter neighbor. It is one of the most ancient open clusters at an age of about 2.5 billion years. Since open clusters are only loosely bound by their mutual gravity, many dissipate when they are much younger. However, M67 lies unusually high above the plane of our galaxy, saving it from the disruptive influences most open clusters face.

M67 is faintly visible through a finder and has even been spotted with the unaided eye under dark skies. Look for it 1.7° due west of Alpha (α) Cancri. A 2.4-inch scope will show just a sprinkling of faint stars over a nebulous haze. My 4.1-inch refractor at 68× re-

On this chart (and all photographs) north is up and east is left. To find north through your eyepiece, nudge your telescope toward Polaris; new sky enters the view from the north edge. (If you're using a right-angle star diagonal, it probably gives a mirror image. Take it out to see an image that matches the map.)

Crabby Delights

Object	Type	Mag	Dist. (l-y)	RA	Dec.
ι¹ Cnc	Double star	4.0, 6.6	300	8ʰ 46.6ᵐ	+28° 45′
γ Cnc	Double star	4.7, 8.7	160	8ʰ 43.3ᵐ	+21° 28′
δ Cnc	Double star	3.8, 11.9	136	8ʰ 44.7ᵐ	+18° 09′
M44	Open cluster	3.1	520	8ʰ 40.0ᵐ	+19° 58′
ADS 6915	Triple star	6.7, 7.3, 7.5	520	8ʰ 39.9ᵐ	+19° 33′
ADS 6921	Quadruple star	6.4, 7.6, 9.2, 10.4	520	8ʰ 40.4ᵐ	+19° 40′
M67	Open cluster	6.9	2,600	8ʰ 50.5ᵐ	+11° 49′

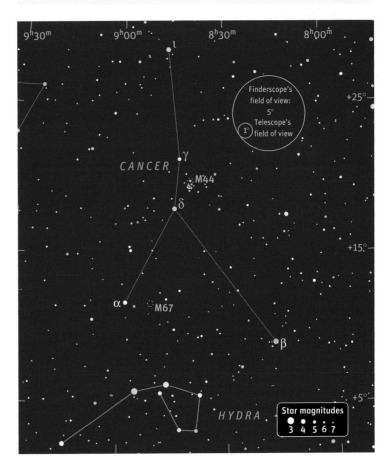

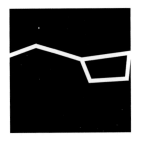   

Chapter

2

Springtime is galaxy time. Why not explore some of the members of the great Virgo Cluster; about 50 of these galaxies are bright enough to be visible in a small telescope. Don't overlook several other galaxies in Ursa Major, Boötes, and Leo. But in case your sky isn't dark enough to hunt galaxies, I've also included more than 20 pretty double stars scattered throughout the constellations of spring.

Stalking the Leo Triplet

LEO, THE LION, contains a distinctive, sickle-shaped asterism that makes the constellation easy to recognize. You can see the Sickle of Leo placed squarely in the middle (just below the zenith) of the all-sky map on page 23. The bright star Regulus (α) Leonis) marks the Lion's heart, while the curve of stars above outlines his sideways-facing head and mane.

To the east of the Sickle, you'll see three stars at the corners of a right triangle. The brightest is Beta (β) Leonis, or Denebola, whose name is abbreviated from the Arabic *Al Danab al Asad* — The Tail of the Lion. The other two stars, Delta (δ) and Theta (θ) Leonis, form the Lion's hindquarters.

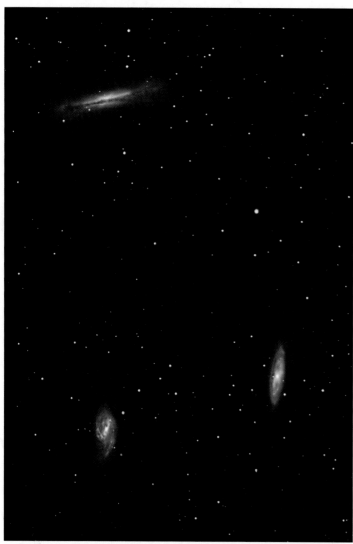

The galaxy triplet of M65, M66 (lower right and left, respectively), and NGC 3628. All three are spirals, with 3628 seen edge on. The field shown here is as wide as the full Moon.

A small telescope will show the M66 Group, a pretty trio of interacting spiral galaxies near Theta Leonis. To locate them, use your lowest-power eyepiece. Sweep 2.2° south from 3rd-magnitude Theta to the 5th-magnitude, orange star 73 Leonis. Then scan 0.8° east, where you'll find a lone 7th-magnitude star. If your eyepiece gives a true field of 50′ or more, you can fit this star and the three galaxies within the same view.

Let's start with **M65,** 18′ south of our 7th-magnitude star. It is faintly visible in a 50-millimeter finder or binoculars. Through my 4.1-inch (105-millimeter) refractor at 47×, I see a fairly bright, oval core with fainter extensions that give the galaxy a very gentle S-curve. At 87×, the galaxy appears to be elongated by a ratio of about 5:1, running nearly north-south. Also at that power, M65 seems to lose its curve, and you can spot a faint star gleaming along the galaxy's southwestern edge. Detailed images show that one of its spiral arms is slightly displaced. This galaxy has a warped disk, probably from gravitational interactions with its companions.

Our next galaxy, **M66,** lies 20′ east-southeast of M65. It looks brighter than M65 to me but leaves some skygazers with the opposite impression. The total integrated magnitude (how bright an object would appear if its light were gathered into a single, starlike point) of M66 is 8.9, while that of M65 is 9.3. But a galaxy is not a starlike point. Its light is spread out over a measurable area of the sky, and the more it is spread the dimmer it will look. As a result, M65 and M66 have nearly the same surface brightness, no doubt contributing to the conflicting opinions of observers. Which do you find easier to see?

Through my 4.1-inch scope at 47×, M66 shows a stellar nucleus, a bright, oval core, and a fainter, oval halo. The oval runs north-northwest to south-southeast. At 87×, M66 is twice as long as it is wide. A 10th-magnitude star can be seen near the northwestern edge of the galaxy.

M65 and M66 were discovered by Pierre Méchain in 1780 and reobserved by Charles Messier for inclusion in his catalog. Messier discovered a comet in October 1773 that passed between these two galaxies the following month. In his catalog, Messier commented that the light of this comet must have kept him from seeing these nebulae. All of Messier's galaxies were referred to as nebulae, and their true nature remained unknown until 1924. The spiral arms of M66 are greatly distorted, earning M66 a listing as Arp 16 in astronomer Halton Arp's *Catalogue of Peculiar Galaxies*. This distortion is attributed to a tidal encounter with the third member of the Leo Triplet: **NGC 3628.**

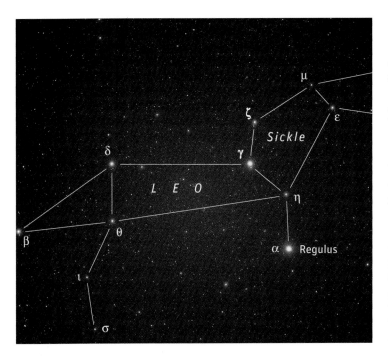

Left: When scanning the sky for Leo, look for the distinctive sickle shape that forms the Lion's head and front quarters. The bright star Regulus (α) lies at the Lion's heart, and Denebola (β) is his tail. *Below:* On this chart (and all photographs) north is up and east is left. To find north through your eyepiece, nudge your telescope toward Polaris; new sky enters the view from the north edge. (If you're using a right-angle star diagonal, it probably gives a mirror image. Take it out to see an image matching the map.) Leo is home to many galaxies. Look for the galaxy triplet containing M65, M66, and NGC 3628 just south of the Lion's hindquarters.

Leo's Galaxy Triplet (M66 Group)

Object	Type	Mag.	Dist. (l-y)	RA	Dec.
M65	Galaxy	9.3	29 million	11ʰ 18.9ᵐ	+13° 05′
M66	Galaxy	8.9	25 million	11ʰ 20.4ᵐ	+12° 59′
NGC 3628	Galaxy	9.5	32 million	11ʰ 20.4ᵐ	+13° 35′

NGC 3628 lies just 36′ north of M66. It has about the same total magnitude as M65, but its light is spread over a larger area, yielding a much lower surface brightness. Despite this, it shows up well in my husband's 3.6-inch refractor under our semirural skies. However, moderate light pollution or haze can easily wipe it out through a small telescope.

My 4.1-inch refractor at 87× gives a true field of 53′, nicely encompassing all three galaxies. This sight is all the more memorable when you realize that each faint smudge is a collection of hundreds of billions of stars. NGC 3628 appears very long and thin, brightening gradually to a slightly brighter, elongated core. The needle shape of this galaxy runs east-southeast to west-northwest, and it appears about eight times longer than it is wide. NGC 3628 is a spiral galaxy whose disk is seen edge on, making it one of the prettiest "flat galaxies" in the sky.

On some professional photographs and deep CCD images, NGC 3628 shows a large plume of material extending 40′ to the east. At a presumed distance of 32 million light-years, the plume would be an incredible 400,000 light-years long. It is thought that the tail was drawn out during a close encounter with M66 about 800 million years ago.

In the 1950s Swiss-born astronomer Fritz Zwicky suggested that such galaxy tails might form star clusters or dwarf galaxies. The tidal plume in NGC 3628 does indeed contain clumpy regions of intense star formation — each comprising a few million solar masses. As the plume disperses over time these groups of stars may be freed to pursue their own course around the galaxy.

The core of NGC 3628 is also undergoing great turmoil. The unusually high rate of star formation around its nucleus classifies this object as a *starburst galaxy.* Starburst galaxies are often members of interacting pairs or groups, where gravitational perturbations

bring large amounts of gas and dust to their central regions. The most active starbursters could decimate their supply of gas in just a few million years (if current rates of star formation continue), while supernovae from the most massive, short-lived stars would drive out the tattered remains. (A well-known starburst galaxy is M82; see page 74.) Only the vast number of galaxies in the universe allows us to catch a glimpse of some of them during such fleeting moments of their astronomically long lives.

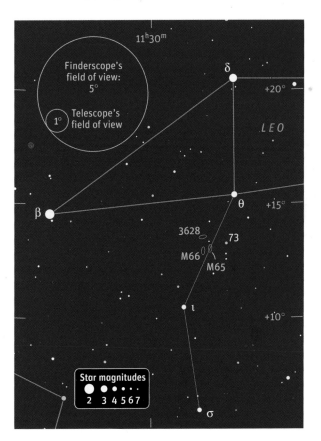

Gregarious Galaxies

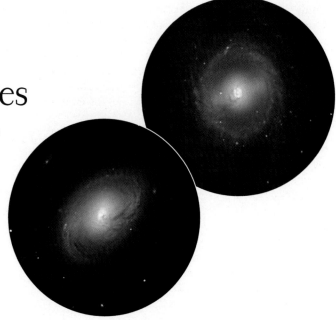

GALAXIES ARE SOCIAL CREATURES. Most gather into groups that range in size from exclusive cliques with just a few individuals to vast flocks of several thousand. The relatively small M96 Group, or Leo I, is well placed for observing on spring evenings. This is the nearest galaxy cluster that contains bright spirals and a bright elliptical galaxy. The M96 Group may be physically associated with the nearby Leo Triplet (or M66 Group), which was featured on the previous page.

First let's visit our cluster's namesake galaxy, **M96.** On the all-sky star chart on page 23 you'll see Leo's brightest star, Regulus, near the center of the chart (just below the zenith) and also quite close to the ecliptic. If you look a little eastward along the ecliptic you will see a much dimmer blue-white star. That's Rho (ρ) Leonis; see the finder chart at right. M96 lies about one-third of the way from Rho to Theta (θ) Leonis (the star at the right angle of the triangle that defines Leo's hindquarters).

For more accurate pointing, use your finder and the chart opposite to locate 5.3-magnitude 53 Leonis 4.2° east-northeast of Rho; it is the brightest star in the area and forms a nice little equilateral triangle with a 7th- and an 8th-magnitude star nearby. Switching to your telescope and a low-power eyepiece, scan 1.4° north-

Taken with a Meade 16-inch telescope, these images bring out structural differences in the otherwise similar pair. M95 *(right)* has a pronounced central bar with a spiral arm coming off each end. M96 *(left)* is notable for the twisting dust lane. Telescopes seldom show these features when used visually.

northwest of 53 Leonis to look for the 9th-magnitude blur of M96. Through my 14 × 70 binoculars, M96 is visible as a faint, fuzzy patch with some brightening toward the center. My little 4.1-inch refractor at 127× reveals only a little more detail. The galaxy is slightly oval, and the long dimension runs northwest to southeast. Its dim outer halo contains a large, brighter core and a stellar nucleus.

In 1998 an Italian amateur astronomer, Mirko Villi, discovered a Type Ia supernova in M96. These supernovae are considered good "standard candles" for determining the distance to remote galaxies. Such an event in a nearby galaxy — one whose distance had already been determined by other techniques — sparked intense interest among professional astronomers. Several recent studies, using improved values derived from this stellar outburst, have helped astronomers adjust the cosmic distance scale and refine the rate of expansion of the universe. The most recent work on Villi's star (designated Supernova 1998bu) yields a rate of about 70 kilometers per second per megaparsec. This is remarkably close to the weighted value of 72 found in the Hubble Space Telescope Key Project. For currently favored cosmological models, either one implies that the universe's age is about 13 billion years.

Two other Messier galaxies belong to the M96 group. **M95** lies 42′ west of M96, and the two will fit together in the same low-power field. Through 14 × 70 binoculars, M95 is a very dim smudge. It has the lowest visual mag-

The Moon could barely fit in this view without covering the galaxies M95 (right) and M96. That means a low-power, wide-field eyepiece (and no Moon!) will show both objects at once. George R. Viscome shot them using an 8-inch f/5.6 reflector and a 45-minute exposure.

nitude and surface brightness of the Messier trio. My 4.1-inch scope at 127× shows a faint, roundish halo and a brighter core that intensifies toward a tiny bright nucleus. With his 4-inch refractor, noted observer and *Sky & Telescope* contributing editor Stephen James O'Meara has been able to detect the little wings that extend beyond the core of this barred spiral as well as a faint outer ring running around the galaxy's edge.

The elliptical galaxy **M105** is located 48' north-northeast of M96. M105 has about the same visual magnitude as M96 but is smaller and therefore has a higher surface brightness. Through my 70-mm binoculars, this galaxy is small and faint and shares the field with M95, M96, and 53 Leonis. The four objects are arranged in a **Y** shape. A small telescope at 30× can encircle all three simultaneously.

In my 4.1-inch scope at 87×, I can see M105 and another galaxy, **NGC 3384,** just to its east-northeast. Their centers are a mere 8' apart. On M105, the same instrument at medium to high power shows a tiny, bright nucleus embedded in a slightly oval glow that fades gradually toward the periphery. NGC 3384 looks like a smaller, dimmer version of M105. Close inspection reveals a very faint third galaxy, **NGC 3389,** only 11' east-southeast of M105. NGC 3384 is a member of the Leo I Group, but NGC 3389 is generally thought to be a background galaxy.

French observer Pierre Méchain was first to notice the three main galaxies of the M96 Group in 1781, but he did not pass on his discovery of M105 in time for inclusion in Charles Messier's final catalog. M105 was added to the list much later, by Helen Sawyer Hogg in 1947.

At least three other Leo I galaxies are within the grasp of a small telescope, all lying within a finder field of M105. Each appears brighter than nonmember NGC 3389 but dimmer than NGC 3384. Search for **NGC 3377** 1.4° north of M105 and 23' southeast of 5.5-magnitude, yellowish 52 Leonis. It is small, not quite round, and has a tiny, bright nucleus. **NGC 3412** lies 1.1° northeast of M105 and 16' southwest of a pair of white and golden 8th-magnitude stars. It is faint, very small, and roundish with a

brighter, stellar nucleus. **NGC 3489** is the most difficult of the three to aim at. There are no bright stars or distinctive pairs to serve as guideposts, so use the chart below to star-hop along the snaking chain of 8th- and 9th-magnitude stars that begins northeast of the M105 galaxy triplet. Alternatively, you can put the white and golden pair of stars near NGC 3412 in the southern part of a low-power field and then scan slowly eastward looking for the telltale glow of NGC 3489. It is a small oval comparable in brightness to NGC 3412, and it has a brighter nucleus.

The galaxies of Leo I are near enough to our own Milky Way that astronomers can study many of the Cepheid variable stars they contain. These well-known distance indicators have placed the cluster at about 38 million light-years, roughly two-thirds the distance to the great Virgo Galaxy Cluster (see page 78).

Deep-Sky Denizens of the M96 Group (Leo I)

Object	Galaxy Type*	Magnitude	Size	RA	Dec.
M96	Sbp	9.3	7.1' × 5.1'	10ʰ 46.8ᵐ	+11° 49'
M95	SBb	9.7	7.4' × 5.1'	10ʰ 44.0ᵐ	+11° 42'
M105	E1	9.3	4.5' × 4.0'	10ʰ 47.8ᵐ	+12° 35'
NGC 3384	E7	9.9	5.9' × 2.6'	10ʰ 48.3ᵐ	+12° 38'
NGC 3389	Sc	11.9	2.8' × 1.3'	10ʰ 48.5ᵐ	+12° 32'
NGC 3377	E5	10.4	4.4' × 2.7'	10ʰ 47.7ᵐ	+13° 59'
NGC 3412	E5	10.5	3.6' × 2.0'	10ʰ 50.9ᵐ	+13° 25'
NGC 3489	E6	10.3	3.7' × 2.1'	11ʰ 00.3ᵐ	+13° 54'

*Galaxies belong to two broad categories, those of spiral form (S) and the less-structured ellipticals (E).

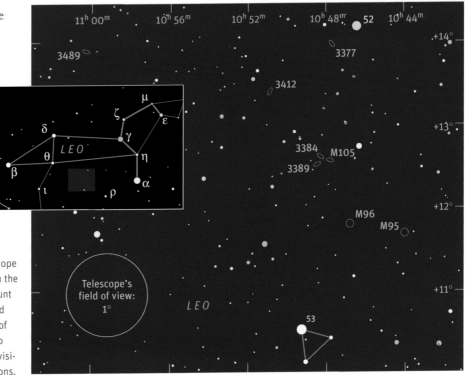

With the help of the inset chart, aim your telescope so the stars 52 and 53 Leonis are both visible in the finder. Once that's been done you're ready to hunt down the galaxies listed above, keeping in mind that the main scope has a much narrower field of view (indicated at lower left). This chart goes to 11th magnitude and shows nearly all the stars visible in a 3-inch telescope under the best conditions.

The Sickle's Harvest

ON OUR APRIL all-sky chart on page 23, Leo, the Lion, holds sway high in the south. Just below the zenith marker is Leo's distinctive Sickle asterism, which resembles a backward question mark. Leo is one of the oldest constellations; its genesis may reach back more than six millennia. Perhaps surprising is that the Sickle, too, seems to have ancient roots.

The Sickle harbored the Sun during summer-solstice grain-cutting time in early Mesopotamia, where the star pattern came to be known as Gis-mes, the Curved Weapon. The Sogdians (originating from areas now part of Uzbekistan and Tajikistan) saw this asterism as

Regulus (bottom center) outshines the feeble glow of the dwarf galaxy Leo I (top center) by 3,000 times, but the latter is 10,000 times farther away. They lie just ⅓° apart on the sky.

a scimitar, calling it Khamshish. According to star-lore expert Richard H. Allen, the Roman encyclopedist Pliny is thought to have included the Sickle in his list of constellations. And in an extinct Celtic language, Manx, the constellation Leo was Yn Corran, the Sickle.

The Sickle plays home to some well-known deep-sky objects that range from quite easy to extremely difficult. Let's start with one of the former.

The tip of the Sickle's blade is marked by a yellow star, 3rd-magnitude Epsilon (ε) Leonis. From there, look for orange, 4th-magnitude Lambda (λ) 3.3° to the west-southwest. A straight drop 1.4° south from Lambda will bring us to our first deep-sky treasure, **NGC 2903,** a barred spiral galaxy similar in size and shape to our Milky Way.

NGC 2903 is faintly visible through a 50-millimeter finder. With my 4.1-inch (105-mm) refractor at 87×, this is a very nice galaxy about 8′ by 4′ in angular size. The outer halo is faint, but there is considerable brightening toward the center. Some dark patches in the oval core hint at spiral structure, and a stellar spot can be seen south of center. A bright patch in the galaxy's northeastern spiral arm bears its own designation, **NGC 2905,** since this looked like a double nebula to its discoverer, William Herschel, who observed from southern England in the late 18th century. You'll probably need at least an 8-inch scope to catch sight of NGC 2905.

Now jump to the easternmost star of the Sickle, **Gamma (γ) Leonis.** It is also known as Algieba, an Arabic-sounding name that may have been coined from *juba,* the Latin word for a lion's mane. Algieba is a lovely double star for a small telescope. My little refractor at 147× shows a close pair of yellow suns, the 3.6-magnitude secondary lying southeast of the 2.4-magnitude primary. Other skygazers have described different colors such as orange and yellow, yellow and green, or even white and pale red. How do they appear to you? Spectral classes of K0 and G5 suggest that their true colors are yellow-orange and yellow.

One of Leo's most picturesque galaxies, NGC 2903, lies near the Sickle's tip. The knot seen here above the galaxy's center, in one of its inner spiral arms, is probably what William Herschel believed to be a second, fainter object. The 7th-magnitude star in the upper-right corner actually lies due north of the galaxy by about ⅓°. This image was taken by Bobby Middleton with a 12½-inch reflector.

William Herschel discovered this double in 1782. Since it has a period of around 600 years, less than half its orbital path has been witnessed by all the ensuing generations of astronomers. Both components are giants that have exhausted the supply of hydrogen fuel in their cores. The primary is 180 times more luminous than our Sun and 23 times larger in diameter, while the corresponding figures for the secondary are 50 and 10.

The U.S. Naval Observatory's *Washington Double Star Catalog* lists two additional companions for Algieba, but their motions show that they are not true members of the system. Both are 10th magnitude and lie more than 5′ to the west-northwest.

Next we'll move to the end of the Sickle's handle, marked by Regulus, Leo's brightest star. The dwarf galaxy **Leo I** sits just 1/3° north of Regulus, whose glare makes the galaxy difficult to see. Leo I is large, but it has very low surface brightness. Besides being a member of the Local Group, Leo I may be the most distant of our Milky Way's known satellite galaxies.

To make the effort a bit easier, choose an eyepiece that will leave Regulus out of the field. With my refractor at 87×, Leo I is very faint, slightly blotchy, and has a few faint stars superimposed. The orientation appears to be east-northeast to west-southwest with dimensions of about 8′ by 5.5′. When I observed this object with Lew Gramer at the Winter Star Party, Lew thought he could see more of the galaxy beyond the area I described. I tried to detect this for quite a while, finally conceding that it was just a "maybe" for me. If you have trouble spotting Leo I, console yourself with the thought that this dim galaxy wasn't even discovered until 1950.

Our final object will be **R Leonis,** one of the easiest variable stars for a small-scope user to follow. This Mira-type pulsating variable typically ranges from magnitude 5.8 to 10.0, with extremes of 4.4 to 11.3. The average period to complete one cycle is 312 days, so observing it once a week will adequately track its changes.

To locate R, search for golden 18 Leonis 5.4° due west of Regulus. Centering 18 Leonis in a low-power field, you'll see 19 Leonis 21′ southeast. From there, drop 9′ south to a 4′ triangle of stars. The western side is marked by 9th- and 10th-magnitude stars, and reddish R Leonis marks the eastern corner.

Deep-Sky Objects Near Leo's Sickle

Object	Type	Mag.	Size/Sep.	Dist. (l-y)	RA	Dec.
NGC 2903	Galaxy	9.0	12.6′ × 6.6′	25 million	9h 32.1m	+21° 30′
NGC 2905	Part of galaxy	—	~0.5′	25 million	9h 32.2m	+21° 31′
γ Leo AB	Double star	2.4, 3.6	4.6″	126	10h 20.0m	+19° 51′
γ Leo AC	Optical double	2.4, 9.6	323″	—	10h 19.6m	+19° 52′
γ Leo AD	Optical double	2.4, 10.0	362″	—	10h 19.6m	+19° 54′
Leo I	Galaxy	9.8	9.8′ × 7.4′	850,000	10h 08.5m	+12° 18′
R Leo	Variable star	4.4–11.3	—	330	9h 47.6m	+11° 26′

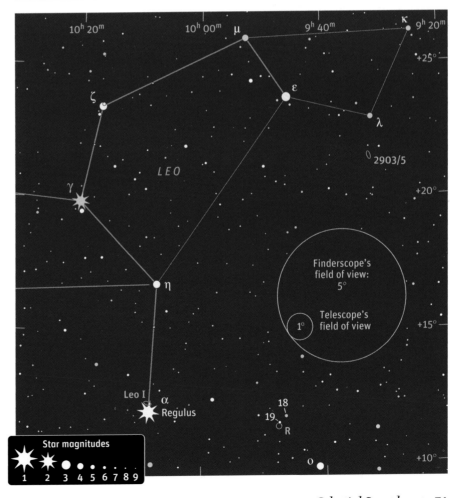

R Leonis's brightness changes can be followed using the chart above, which gives the magnitudes of surrounding stars (numbers in italics, with the decimal point omitted). To locate this telescopic field, first find R Leonis on the large chart at right.

Highlights of Hydra

HYDRA, THE WATER SNAKE, is our largest constellation both in area and in length. The she-serpent winds across more than a quarter of the heavens, and this is the only time of year when the evening sky will show you Hydra in her entirety. Her head is marked by an oval of stars south of Cancer, the Crab, while her tail abuts Libra, the Scales. With her enormous length, Hydra can keep you entertained throughout the evening as the passing hours bring each new wonder to its highest place in the sky.

Our first showpiece is **M48,** the brightest open star cluster in Hydra. As twilight ends in April the cluster is already descending toward the west, so you should catch this one early in the evening. To locate M48, look for 3.9-magnitude C Hydrae 8° south-southwest of Hydra's head. C Hydrae is easy to recognize in a finder because two 5.6-magnitude stars flank it. Placing C near the northeastern edge of your finder's field will bring the hazy glow of M48 into view.

When you switch to a low-power telescopic view, M48 is a beautiful sight! My 4.1-inch scope at 47× shows more than a hundred stars of 8th through 12th magnitude in nearly 1°. The central 1/2° is more densely populated, while the stars loosely cast around the periphery blend into the background sky. The group's brightest member is found at the southeastern edge of the central mass and shines with a yellowish hue.

M48 was once considered a missing Messier object, since there is no cluster at the coordinates Charles Messier gave in the late 18th century. Some older atlases plot M48 in the original, incorrect position — a fact I discovered the hard way when learning my way around the

Unlike a close-up image with a large telescope, this wide-field photograph of the galaxy M83 more nearly shows what observers can expect to see with small telescopes. This view, 3° wide, includes three stars on the finder chart opposite; north is up.

sky. Messier described M48 as a cluster of faint stars near the three stars that begin the Unicorn's tail. These stars are C Hydrae and its attendants, which no longer belong to the Unicorn (Monoceros), according to modern constellation boundaries. Since the cluster NGC 2548 fits Messier's description, newer atlases equate it with M48.

Our next target will be Hydra's brightest planetary nebula, **NGC 3242,** or Caldwell 59, popularly known as the Ghost of Jupiter. In 1785 William Herschel discovered this nebula and was the first to compare it to that planet, saying that its light was the color of Jupiter. The nickname was coined by William Noble. In his 1887 book *Hours With a Three-Inch Telescope* he wrote: "It will be seen as a pale blue disc, looking just like the ghost of Jupiter."

NGC 3242 is found 1.8° south and slightly west of 3.8-magnitude Mu (μ) Hydrae. The planetary is easy to see through a finder but looks like an 8th-magnitude star. You might be able to recognize it by its non-starlike color, which is usually described as blue or green. Magnifications of 50× or more will reveal its true nature.

Despite Herschel's claim, the color of NGC 3242 does not resemble Jupiter's to my eyes. With my little refractor the planetary looks distinctly turquoise-blue. At 203× it is roundish, nearly uniform in surface brightness, and slightly fuzzy around the edge. Those with better skies or larger telescopes should look for a northwest-to-southeast elongation with brighter patches at each end. You might also be able to spot the 12th-magnitude central star.

Now we'll move to the Water Snake's reddest star, **V Hydrae.** V Hydrae is a carbon star and, as such, has a number of carbon-bearing molecules in its atmosphere that act like a red filter. It varies in a period of roughly 17 months superposed on a grander cycle spanning 18 years. The short-term change currently seems to be taking the star from about magnitude 6.5 to 9.5, well within the reach of a small scope.

To find V Hydrae, look first for 5th-magnitude b³ Hydrae. It forms a right triangle with 4th-magnitude Al-

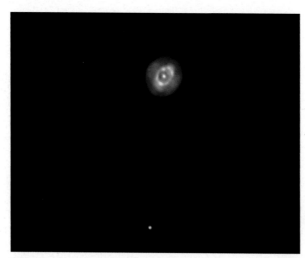

While it looks tiny, pale, and ghostly in small scopes, the planetary nebula NGC 3242 shows fine structure in this CCD image. The 11th-magnitude star at bottom lies just 2.4′ south of the nebula.

pha (α) Crateris and 3rd-magnitude Nu (ν) Hydrae, all visible within the same finder field. A low-power view through your telescope will show that b³ Hydrae forms another right triangle with a 6.6-magnitude star to its southeast and a 7.1-magnitude star to its south. Drawing a gentle curve from b³ Hydrae through the southern star and continuing for a little more than that distance again should bring you to V Hydrae. The star's color varies along with its brightness, from ruddy at minimum light to deep orange at maximum.

V Hydrae appears to be a dying red-giant star ready to form a planetary nebula. Studies indicate that planetary nebulae are largely shaped by high-speed bipolar outflows that cover a mere few hundred to a thousand years of a star's lifetime. V Hydrae is the first star to be caught in the act. A paper in the November 20, 2003, issue of *Nature* reports its capture with the Hubble Space Telescope's imaging spectrograph. The research team involved suggests that the outflow may be driven by an unseen companion star or giant planet.

Hydra's brightest globular cluster is **M68.** The stars of Corvus, the Crow, serve as handy pointers for tracking it down. Draw an imaginary line from Delta (δ) through Beta (β) Corvi and continue for half that distance again. There, through a low-power eyepiece, you will see a 5th-magnitude star with a fuzzy spot to its northeast.

My 4.1-inch scope at 87× turns the fuzzy patch into a 9' ball of light with a large, bright, mottled core surrounded by a sparse and tattered halo that sparkles with faint stars. Increasing the magnification to 153× plucks out more of the cluster's stars right down to the center. The dappled face of M68 prompts some observers to picture dark lanes running through it.

Our final target is the brightest galaxy in Hydra, **M83,** conveniently placed two-thirds of the way from 3rd-magnitude Gamma (γ) Hydrae to 4th-magnitude 1 Centauri. Northern skygazers may find the latter dimmed below naked-eye visibility when the sky is bright or the horizon hazy. If so, try star-hopping south-southeast from Gamma along a curvy line of 6th- and 7th-magnitude stars. M83 lies 1° east-southeast of the last star in the chain and can share the same low-power field.

At 127× my little scope shows a small bright core

and a fairly bright inner halo, 5' by 2', running east-northeast to west-southwest. This is surrounded by a dim oval halo about 8' across. I see some brighter patches but cannot trace M83's spiral structure. Three 10th-magnitude field stars form a tangent to the southeastern side of the galaxy.

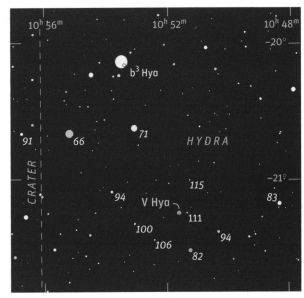

This close-up chart for V Hydrae gives the magnitudes of comparison stars to tenths, with the decimal point omitted. Use the larger map below to find V Hydrae, located near Alpha (α) Crateris.

Celestial Treats in Hydra

Object	Type	Mag.	Size	Dist. (l-y)	RA	Dec.
M48	Open cluster	5.8	54'	2,500	8ʰ 13.7ᵐ	–5° 45'
NGC 3242	Planetary nebula	7.3	25"	2,900	10ʰ 24.8ᵐ	–18° 39'
V Hya	Carbon star	6.5–9.5	—	1,600	10ʰ 51.6ᵐ	–21° 15'
M68	Globular cluster	7.8	11'	33,000	12ʰ 39.5ᵐ	–26° 45'
M83	Spiral galaxy	7.5	13'×11'	15 million	13ʰ 37.0ᵐ	–29° 52'

Angular sizes are from catalogs or photographs. Most objects appear somewhat smaller when a telescope is used visually.

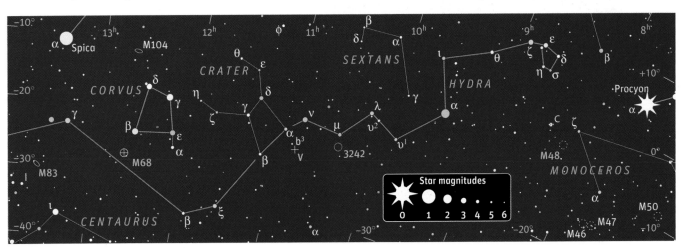

Bear Up!

URSA MAJOR, the Big Bear, is the third-largest constellation in the sky, and its seven brightest stars form the well-known pattern of the Big Dipper, visible all night long from midnorthern latitudes. At this time of year the Bear strides high and upside down across the northern sky during the evening with her toes brushing the zenith. Most stargazers are familiar with using the front two Bowl stars — Merak or Beta (β) Ursae Majoris, and Dubhe or Alpha (α) Ursae Majoris — to point to Polaris, the North Star.

The galaxies M81 and M82 lie just northwest of the bowl of the Big Dipper. The galaxies fit easily into a 1° field of view. I sketched them *(facing page)* while looking through a 4.1-inch f/5.8 refractor at 68×. North is up.

This vast tract of the celestial vault contains an abundance of deep-sky delights, so let's turn our scopes toward Ursa Major and see what we can find.

M81 and M82 look lovely through a small telescope. These two galaxies are the brightest members of the M81 galaxy group at a distance of about 13 million light-years. Look for them behind what many consider the Bear's ear, the 4.5-magnitude star 24 Ursae Majoris. M81 is bright enough to be glimpsed through a good finder or with binoculars. The duo can fit in a 1° field of view with room to spare.

Each galaxy deserves individual study at high pow-

Left: Look southeast of Merak (the Dipper's southern pointer star) for the edge-on spiral galaxy M108 (upper right) and M97, the Owl Nebula (lower left). North is up in this Akira Fujii image.

Below: A close-up of the spiral galaxy M109. Astrophotographer Gérard Therin used a 9¼-inch f/12 Schmidt-Cassegrain telescope, a f/6.3 focal reducer, and a HI-SIS 22 CCD camera to take 21 5-minute exposures, which were combined into the final image. North is to the upper left.

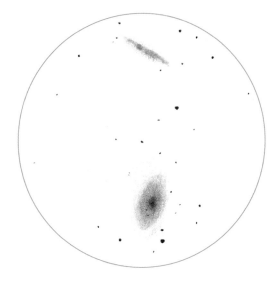

may just see a round, uniform glow. An O III or narrowband light-pollution filter can help combat suburban skyglow when looking for the Owl.

M109 lies just 0.6° east-southeast of Phecda, Gamma (γ) Ursae Majoris. This galaxy is one of the faintest Messier objects and usually needs a 3-inch or larger telescope to be seen. One trick to help you spot it is to use a high enough power to keep bright Phecda out of the field of view. You can also try lightly tapping your telescope's tube to jiggle the images. A faint object seen in motion is easier to detect than a stationary one, especially with averted vision. A 4-inch scope at 100× shows a very faint, oval halo aligned east-northeast to west-southwest surrounding a brighter core.

Ursa Major certainly seems to bear up under scrutiny, and her bright stars make getting our bearings easy. In my next essay we'll visit a much more obscure star pattern and see who's nipping at the Big Bear's heels.

ers. Through my 4.1-inch (105-millimeter) scope, M81 appears oval with a large halo surrounding a bright central region. Two 11th-magnitude stars can be seen in the southern part of the halo. The core brightens toward a nearly stellar-looking nucleus.

Spindle-shaped M82 looks mottled. Its splotchy appearance becomes more evident with increasing aperture, but under dark skies you can detect hints of this even in a 2.4-inch scope. The strange visage of M82 is most likely due to a close encounter with its more massive companion. M81 is thought to have swept by its smaller neighbor long ago; the near collision distorted M82, triggering great bursts of star formation. (See page 67 for more about another starburst galaxy, NGC 3628.)

Dubhe, the end star in the Bowl of the Big Dipper, is a double that is an easy split in binoculars. Golden-hued Dubhe is the only colored star in the Big Dipper. It sports a 7th-magnitude companion 380″ to the south-southwest. Although the dimmer star is yellow-white, it may appear bluish through a small telescope. This contrast effect is frequently seen when we compare a dim star to a brighter yellow or orange one nearby.

You'll find the mottled-looking galaxy **M108** 1.5° east-southeast of Merak. M108 is very faint through a 2.8-inch scope at low power but can be seen to be elongated east-west. A 4-inch scope at moderate to high powers can start to show its disheveled appearance. The galaxy looks like a cigar-shaped collection of light and dark patches with no pronounced core.

Just over 0.8° farther southeast we come to the large, dim planetary nebula **M97**. It forms the eastern corner of a quadrilateral with three 7th- and 8th-magnitude stars. M97 is often called the Owl Nebula because it contains two vague dark markings that resemble large eyes. The eyes are difficult to see in telescopes smaller than 6 inches or in a less-than-dark sky. You

Sights Near the Big Dipper's Bowl

Object	Type	Mag.	Dist. (l-y)	RA	Dec.
M81	Spiral galaxy	7.9	13 million	9ʰ 55.6ᵐ	+69° 04′
M82	Irregular galaxy	9.2	13 million	9ʰ 55.9ᵐ	+69° 41′
Dubhe (α UMa)	Double star	1.8, 7.1	123	11ʰ 03.7ᵐ	+61° 45′
M108	Spiral galaxy	10.7	45 million	11ʰ 11.5ᵐ	+55° 40′
M97	Planetary nebula	10	2,600	11ʰ 14.8ᵐ	+55° 01′
M109	Galaxy	10.5	55 million	11ʰ 57.5ᵐ	+53° 23′

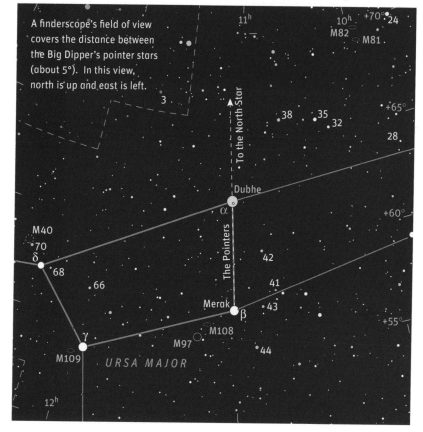

A finderscope's field of view covers the distance between the Big Dipper's pointer stars (about 5°). In this view, north is up and east is left.

Chasing Treasures Under the Dipper

THE SPRING CONSTELLATION Canes Venatici represents the Hunting Dogs, Asterion and Chara. Boötes, the Bear Driver, holds onto their leads as they pursue Ursa Major across the sky. Only two stars in Canes Venatici are bright enough to grace our naked-eye all-sky star chart on page 24, with its limiting magnitude of 4.5. Look for them under the curve of the Big Dipper's handle straight overhead (at the chart's center). Still,

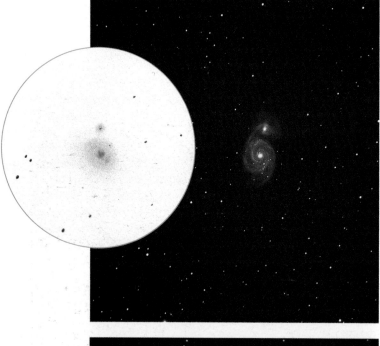

Canes Venatici holds a cosmic treasure trove that belies its lackluster appearance.

Let's start with the brightest star in the constellation, Alpha (α) Canum Venaticorum, popularly known as **Cor Caroli**. When Charles II of England returned from exile to restore the monarchy in 1660, his physician claimed that this star shone with unusual brilliance, as if the heart of the new king's slain father were swelling with joy. English cartographer Francis Lamb added this name to a star chart in 1673 as Cor Caroli Regis Martyris (Heart of the Martyred King Charles).

Cor Caroli is a lovely double star for a small telescope. The 2.9-magnitude primary has a 5.6-magnitude secondary star 19 arcseconds to the southwest. The pair is easily split at 40× and shows a delicate color contrast — the brighter star looks blue-white and the fainter one yellow-white.

If you have trouble seeing the colors in Cor Caroli, take a look at ruddy **Y Canum Venaticorum.** Y CVn is a variable star with a visual magnitude slowly ranging from 5 to 6. Look for it 4.6° north-northeast of Beta (β) CVn, making a right triangle with Beta and Cor Caroli.

Y is deep red not just because it is relatively cool like other red giants. Carbon-containing molecules like C_2, C_3, CN, and SiC in the star's atmosphere absorb much of the blue light from the star, leaving it with an unusually vivid reddish orange color. The beauty of its spectrum earned it the name La Superba from the 19th-century Italian spectroscopy pioneer Pietro Angelo Secchi.

Look for one of the finest globular star clusters in the sky a little more than halfway from Cor Caroli to brilliant Arcturus in Boötes. **M3** contains a half million stars and more known variables than any other globular. It is bright enough to be seen through a finderscope as a tiny, hazy spot $1/2°$ northeast of a 6th-magnitude orange star. A 4-inch telescope at high power begins to reveal faint pinpricks of light sprinkled across a hazy background. The core is slightly elongated northwest-southeast and grows much brighter to-

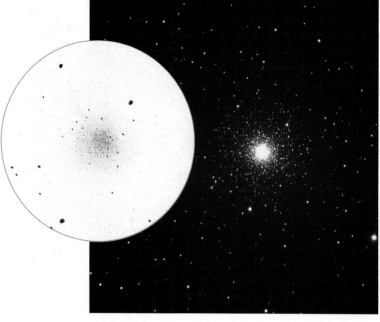

Pair above left: M51 and NGC 5195 lie about 37 million light-years away just off the end of the Dipper's handle. These galaxies can easily fit into a 1° field of view. My sketch shows them through a 4.1-inch f/5.8 telescope at 87×, while Akira Fujii's astrophotograph captures the cool beauty of the pair.

Pair left: The globular cluster M3 contains some of the oldest-known stars. In a 1° field of view, it is easily visible as a hazy spot. I sketched this lovely little cluster using my 4.1-inch f/5.8 refractor at 127×. Akira Fujii photographed M3 through his 12$1/2$-inch f/5 Newtonian telescope, revealing more starry details in this shimmering cluster. North is up in all images.

ward the center. Larger apertures display a dense, glittering ball of snow crystals nestled within a halo of sparkling stars, many arranged in curving chains. These are some of the most ancient stars in our galaxy, born around 12 billion years ago.

A famous spiral galaxy can be found about a quarter of the way from Alkaid, the star at the end of the Big Dipper's handle, to the midpoint between the two brightest stars of Canes Venatici. This is **M51,** the famous Whirlpool Galaxy. In a finderscope its position forms the right angle of an easily remembered asterism with Alkaid and the 6th-magnitude stars 24 and 21 CVn, as shown on the chart ar right.

The galaxy is faintly discernible through a large finder under a dark sky, but any light pollution or atmospheric haze quickly renders it invisible. A 2.4-inch telescope at 50× shows a nearly stellar nucleus surrounded by a small core and a large, very faint halo consisting of the galaxy's spiral arms. The Whirlpool was the first galaxy discovered to have a spiral structure, but the arms are not easy. A 4-inch telescope at 100× can pick out one or two slightly brighter patches in the region of the arms, but it takes at least an 8-inch and a nice, dark night to begin to reveal the true splendor of the Whirlpool's sweeping spiral.

M51's companion galaxy, **NGC 5195,** is smaller but almost as easily noticed. It's the round glow outside the northern edge of M51's halo. A 4-inch at high power will show a tiny, stellar nucleus in NGC 5195 that's about as bright as the nucleus of M51.

The spiral **M63** is sometimes known as the Sunflower Galaxy. It lies nearly halfway from Cor Caroli to M51 and is just 1.2° north of a little asterism of four stars capped by 6th-magnitude 19 CVn. This small oval of light is easily spotted in a good 2.4-inch scope. My 4.1-inch at high power shows a stellar nucleus within a broad, oval core. The core seems elongated east-west, while the fainter surrounding halo runs east-southeast to west-northwest. Through larger instruments the galaxy begins to take on a curdled appearance. M63 is a good example of a multiple-arm spiral galaxy. In photographs we can see that its arms are arranged in many short arcs. Robert Burnham Jr., in his *Celestial Handbook,* likened them to "showers of sparks thrown out by a rotating, fiery pinwheel."

Our brief run with the Hunting Dogs by no means exhausts the riches of this obscure constellation but has merely sampled the types of celestial prey that await the patient deep-sky explorer. The hunting fields are filled with double stars, variables, star clusters, and dozens of galaxies. Tallyho!

Riches Beneath the Big Dipper's Handle

Object	Type	Mag.	Dist. (l-y)	RA	Dec.
Cor Caroli	Double star	2.9, 5.6	110	12^h 56^m	+38° 19′
Y CVn	Variable star	5.1–6.3	710	12^h 45^m	+45° 26′
M3	Globular cluster	6.4	32,000	13^h 42^m	+28° 23′
M51	Spiral galaxy	9.3	37 million	13^h 30^m	+47° 12′
NGC 5195	Galaxy	10.6	37 million	13^h 30^m	+47° 16′
M63	Spiral galaxy	8.6	37 million	13^h 16^m	+42° 02′

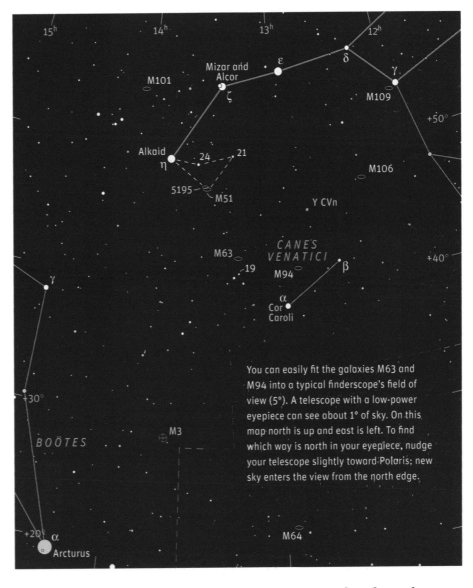

You can easily fit the galaxies M63 and M94 into a typical finderscope's field of view (5°). A telescope with a low-power eyepiece can see about 1° of sky. On this map north is up and east is left. To find which way is north in your eyepiece, nudge your telescope slightly toward Polaris; new sky enters the view from the north edge.

A Toehold in the Virgo Cluster

THE HAZY BAND of the Milky Way runs through the constellations that hug the horizon on May's all-sky chart on page 24. With the plane of our galaxy and its obscuring dust clouds riding so low, we can gaze up into an unobstructed sky and peer far into deep space. Our clearest view is toward the North Galactic Pole in the constellation Coma Berenices, where we just happen to find the nearest large cluster of galaxies.

Some 60 million light-years away, the Virgo Galaxy Cluster splays out across the borders of Virgo and Coma; its location is indicated on the May all-sky map. Its 2,000 member galaxies form the core of the Local Supercluster, with our own Local Group as an outlying member. The tremendous mass of the Virgo Cluster acts on the galaxy groups around it, slowing the recession they would otherwise have as part of the uni-

verse's overall expansion. The Milky Way Galaxy and other members of the Local Group may eventually fall into and be swallowed by the Virgo Cluster.

About 50 Virgo Cluster galaxies, including 16 Messier objects, are bright enough to be seen through a small telescope. With such an abundance of galaxies in an area devoid of bright stars, novice skygazers easily become lost in this realm. In such a case it is wise to remember the old adage: Well begun is half done. We need a toehold in the Virgo Cluster — a familiar base from which to start and then carry our explorations farther afield.

If your sky is dark enough for you to spot the 5.1-magnitude star 6 Comae Berenices, you're already on your way. If not, begin at the bright star Denebola in the tail of Leo, the Lion. Scan eastward from Denebola for 6½° to find 6 Comae. It is the brightest star along the way and the westernmost star in a distinctive T-shaped asterism with four other stars ranging from magnitudes 6.4 to 6.9.

This group of stars will be our home base in the Virgo Cluster. A finder will easily encompass the entire T. The group will also fit within the field of your main scope, if you have a low-power eyepiece that gives you a true field of 2° or more.

First, start at 6 Comae and use it to find the galaxy **M98.** This nearly edge-on spiral lies ½° due west of 6 Comae and shares the field at powers under about 70×. Although M98 has a low surface brightness, it can be seen in a 2.4-inch (60-millimeter) scope under dark skies. Through my 4.1-inch scope at around 100×, the galaxy is about 6′ by 2′, elongated north-northwest to south-southeast. It contains a brighter, extended, patchy core and an off-center, nearly stellar nucleus. Through his 4-inch refractor, noted observer Stephen James O'Meara sees the brighter areas of M98 forming a Klingon vessel from *Star Trek*. Less-experienced observers will probably need a larger telescope to see the arcing spiral arms that give M98 this appearance.

M98 is approaching us at 125 kilometers per second. Since the Virgo Cluster as a whole is receding from us at about 1,100 km/sec, M98 must be moving in our direction at a rate of 1,225 km/sec with respect to the center of its cluster. The immense mass of the Virgo Cluster gravitationally accelerates many of its members to high individual velocities, so that the light from some, like M98, actually shows us an approaching blueshift instead of the more usual redshift of cosmological expansion.

If you have trouble spotting M98, take heart. Our next two targets are a little easier. This time we'll start from the 6.5-magnitude star in the center of our T's upright. The face-on spiral galaxy **M99** lies just 10′ southwest of this star and will even fit in the same high-pow-

Left: The most nearly edge-on of the galaxies discussed here, M98 is also a little fainter than the others. The star 6 Comae Berenices lies just outside the left edge of this close-up view. *Below:* M99 completes a triangle with stars of 6th and 9th magnitude that shine brightly in the same telescopic field. This galaxy's miniature pinwheel, so striking on photographs, requires a large scope to be detected visually.

The symmetry of M100's spiral arms makes it a showpiece. In long-exposure photographs, numerous faint galaxies can be seen in M100's vicinity. But NGC 4312, the needle-like object ⅓° to the south-south-east, is beyond the reach of a small scope visually.

To locate our next galaxy, we'll start at the 6.5-magnitude star at the eastern side of our **T**'s crossbar. The face-on spiral galaxy **M100** lies just 35′ east-northeast of this star. Through my 4.1-inch scope at around 100×, this galaxy appears slightly oval, about 4′ by 3′, and is tipped east-southeast to west-southwest. Its halo is fairly uniform and contains a small, round, brighter nucleus. A 6-inch scope can start to show, within the halo, slightly brighter patches that define the galaxy's spiral arms. Photographs of M100 reveal the beautiful structure that astronomers label a *grand-design spiral* (one having two principal arms).

The **T** of stars and the galaxies M98, M99, and M100 give us our toehold in the Virgo Cluster, but don't stop here. Once you've familiarized yourself with the area, use a good atlas to star-hop, or even galaxy-hop, to new sights. For example, from M100 you can scan 45′ east to a 6.7-magnitude star. From there it is 2.3° north to **M85,** or 2.9° south to **M86.** M86 lies right next to **M84,** and this pair marks one end of a long, distinctive curve of galaxies known as Markarian's Chain (see page 84).

Take it a little at a time and eventually you'll master the brighter galaxies of the Virgo Cluster. Then, someday soon, perhaps you'll be the one showing a novice how to navigate them.

er field of view. My 4.1-inch scope at around 100× shows a 3′-by-2′ oval core that brightens toward the center and is surrounded by a very faint, oval halo. The core's elongation runs east to west, while that of the halo runs northeast to southwest. Under dark skies a 6-inch scope can start to show hints of spiral structure. Knots of slightly brighter haze emerge from the eastern side of the galaxy and wrap around the south.

M99's spiral structure is asymmetric, probably from arm-wrenching gravitational interactions with other Virgo Cluster members. The galaxy is sometimes nicknamed the Pinwheel, a name it shares with two other Messier objects: M33 and M101. M99 also has a high velocity of its own. Its measured redshift, unusually large for a galaxy at M99's distance, indicates it is hurtling away from us at 2,324 km/sec. That's twice the recession rate of the Virgo Cluster as a whole!

Easy Galaxies in the Virgo Cluster

Galaxy	Type	Mag.	Size	RA	Dec.	Constellation
M98	Spiral	10.1	9′ × 3′	12^h 13.8^m	+14° 54′	Coma Berenices
M99	Spiral	9.8	5′	12^h 18.8^m	+14° 25′	Coma Berenices
M100	Spiral	9.4	7′ × 6′	12^h 22.9^m	+15° 49′	Coma Berenices
M85	Elliptical	9.2	7′ × 5′	12^h 25.4^m	+18° 11′	Coma Berenices
M86	Elliptical	9.2	7′ × 6′	12^h 26.2^m	+12° 57′	Virgo
M84	Elliptical	9.3	5′ × 4′	12^h 25.1^m	+12° 53′	Virgo

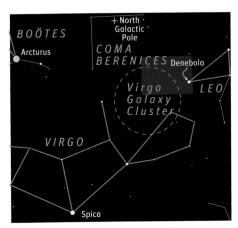

In this essay I touch on just a few of the many marvels in the mighty Virgo Galaxy Cluster. As a whole the cluster spans more than 10° of sky (see inset above), but a good starting place is the T-shaped asterism marked on the close-up chart at right, not far east of Leo's tail star, Denebola.

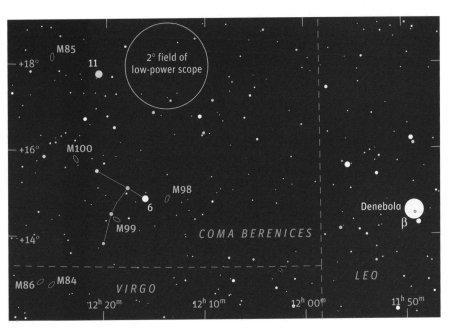

Coma Squared

COMA BERENICES (Berenice's Hair) flows high across the evening skies of spring for observers at midnorthern latitudes. This area is spangled with faint stars when seen under a very dark sky. But only three of these stars are bright enough to make our all-sky map on page 24, where they look like a rotated **L** just above Virgo. Filling in the blank corner of Coma with the galaxy M85 gives us a square 10° on each side, about as wide as your fist at arm's length. In and near this celestial square, deep-sky sights abound.

The constellation Coma Berenices is named for a real person, Queen Berenice II, who lived in the 3rd century BC and was the wife of Ptolemy III, king of Egypt. Shortly after their marriage, Ptolemy went off to war in response to diplomatic intrigues that included the murder of his sister (also named Berenice).

Out of fear for her husband, her king, and (no doubt) her position, Queen Berenice offered her royal tresses to the gods as ransom to secure Ptolemy's safety. When he returned victorious, Berenice made good her vow. Her amber locks were shorn and placed in the temple of Aphrodite, but they vanished shortly thereafter.

The royal couple's outrage was quelled by the palace astronomer, Conon of Samos. Conon wisely claimed that the gods were so pleased with Berenice's lustrous gift that they placed it in the sky for all to see, now

It's no mystery how M64, the Black Eye Galaxy, got its name. The photographic image, taken with a Celestron 11, spans about ¼°. My pencil drawing *(facing page)*, made at 127× with my 4.1-inch refractor, shows a much wider field. Both views are north up.

shining as a delicate tracery of glittering stars:

> [He] who scanned the vast heaven's lights,
> Who watched the risings and settings of stars, . . .
> That same Conon noticed me, Berenice's Lock,
> In my celestial abode.*

We'll start our sky tour at the constellation's Alpha (α) star, which is sometimes called Diadem but is always pictured at the wrong end of the Queen's hair to mark her jeweled crown. The globular cluster **M53** lies 1° northeast of this pale topaz gem. M53 looks nearly stellar through a small finder, but 14 × 70 binoculars reveal a moderately bright, round, fuzzy patch that brightens toward the center. With my 4.1-inch (105-millimeter) refractor M53 begins to look grainy at 153×. At 203×, I can see a few elusive pinpricks of light on a steady night.

If M53 seems easy, try for the very dim globular **NGC 5053** just 1° to the east-southeast. Stephen James O'Meara, who has glimpsed this in his 4-inch refractor, delightfully describes NGC 5053 in *Deep Sky Companions: The Messier Objects* as "the departed soul of its more brilliant neighbor."

Now let's move to the diagonally opposite corner of the Coma square marked by golden Gamma (γ). Scattered over a region 5° across, with Gamma at the northern edge, is the nearby open cluster known as the Coma Star Cluster, or **Melotte 111.** Because this grouping is so large, a regular telescope is of little use here; binoculars or a finderscope work much better. In 8 × 40 binoculars I can count a dozen bright stars and twice as many fainter ones. The cluster includes the very wide, bright double star **17 Comae Berenices.** A small scope at low power shows a subtle color contrast between the 5.2-magnitude primary and the 6.6-magnitude secondary. They are actually blue-white and white, re-

This amazing sliver of light — the almost perfectly edge-on spiral galaxy NGC 4565 — is a fairly easy catch for any small telescope under dark skies. But the tiny smudge ¼° toward lower right, the galaxy NGC 4562, is a tough challenge visually for 12-inch telescopes.

---

*From poem 66 of the Roman lyric poet Catullus, who lived in the 1st century BC. It is full of astronomical allusions.

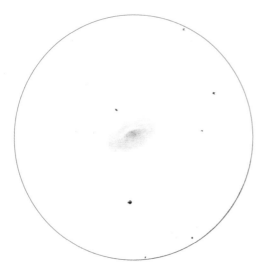

galaxy **M85.** If your sky is dark enough, you can try to spot nearby 4.7-magnitude 11 Comae with the unaided eye (see the finder chart below). It looks yellow-orange through the eyepiece, and you'll find M85 only 1.2° (one low-power field) to the east-northeast. M85 is the northernmost Messier galaxy in the Virgo Galaxy Cluster and one of the brightest. It is easily visible in my 14 × 70 binoculars as a small, oval patch of light. My little refractor at 87× shows a 2′ tapered oval that grows smoothly brighter toward the center. A very faint star can be seen on the northern edge, and a 10th-magnitude star lies close to the southeast. In dark skies it's possible to glimpse the much fainter galaxy **NGC 4394** east of M85 — their centers are separated by a mere 7.5′. It appears as a small oval with a bright, stellar nucleus.

spectively, but some observers see the companion star as blue-green.

NGC 4565 (Caldwell 38) lies 1.7° east of 17 Comae. It is one of the most impressive galaxies in the *Revised Flat Galaxy Catalogue* (I. D. Karachentsev and colleagues, 1999), which lists thousands of highly elongated, edge-on spirals. My little refractor at 47× shows NGC 4565 about 7′ long and very thin with a small, brighter bulge in the center. Switching to 87× brings out some mottling across the core, hinting at the dark lane that runs along the galaxy's length.

Another nice galaxy can be seen 2° north of NGC 4565. **NGC 4559** (Caldwell 36) is also elongated, but it appears a little shorter and fatter than its neighbor to the south. Two faint stars lie on either side of the galaxy's southeastern end.

Nearly two-thirds of the way from 17 Comae to Alpha we find the 4.9-magnitude golden star **35 Comae.** It is easy to recognize as the brightest star in the area, and it has a 9.8-magnitude companion to the southeast that is well separated at 17×.

We can use 35 Comae to find **M64,** the Black Eye Galaxy, 1° to the east-northeast. The Black Eye gets its name from the prominent dust lane that looks so stunning in photographs. Observers have been able to glimpse this feature in telescopes as small as 60 mm. In my 4.1-inch refractor at 127×, the galaxy appears about 6′ by 3′.

Now we'll move to the southwestern corner of Coma's square, marked by the

Zooming in on the three main stars of Coma Berenices, this chart includes fainter stars to magnitude 9.5 for help in locating the objects in this essay. Most finderscopes have a field as wide as the dotted circle at top right, which encloses the loose star cluster Melotte 111.

Deep-Sky Riches of Coma Berenices

Object	Type	Mag.	Size/Sep.	Dist. (l-y)	RA	Dec.
M53	Globular cluster	7.6	12′	60,000	13h 12.9m	+18° 10′
NGC 5053	Globular cluster	9.5	8′	54,000	13h 16.5m	+17° 42′
Mel 111	Open cluster	1.8	5°	290	12h 25m	+26.1°
17 Com	Double star	5.2, 6.6	145″	270	12h 28.9m	+25° 55′
NGC 4565	Galaxy	9.6	16′ × 3′	32 million	12h 36.3m	+25° 59′
NGC 4559	Galaxy	10.0	11′ × 5′	32 million	12h 36.0m	+27° 58′
35 Com	Double star	5.0, 9.8	29″	320	12h 53.3m	+21° 15′
M64	Galaxy	8.5	9′ × 5′	19 million	12h 56.7m	+21° 41′
M85	Galaxy	9.1	7′ × 5′	60 million	12h 25.4m	+18° 11′
NGC 4394	Galaxy	10.9	3′	60 million	12h 25.9m	+18° 13′

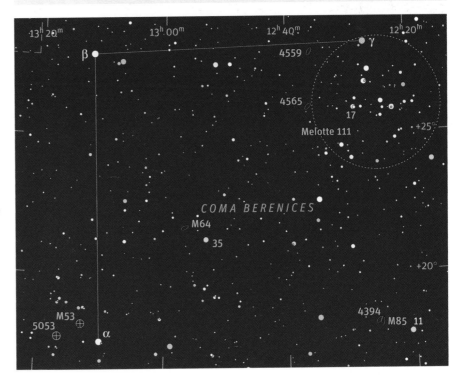

Under Virgo's Wing

VIRGO, THE MAIDEN, is portrayed with wings on many pictorial star atlases, including Johann Bayer's unprecedented *Uranometria* (Augsburg, 1603) and Johann Bode's magnificent *Uranographia* (Berlin, 1801). Virgo's wings may spring from association with Dike, or Astraea, goddess of human justice in Greco-Roman mythology. She originally dwelt on Earth during the Golden Age of perfect peace and happiness. In the ensuing Silver Age people became impious and quarrelsome, and the goddess, rebuking them and warning of worse to come, withdrew to the mountains. Mortals grew more wicked during the Bronze and Iron Ages, waging war and abandoning justice, so the goddess abandoned Earth and flew up into the heavens, where we see her wearing the wings that bore her away.

Virgo's northern wing is adorned with a host of galaxies bright enough to be enjoyed with a small tele-

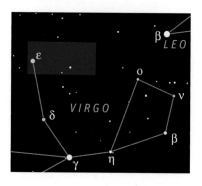

In a simple star-hop westward from the 2.8-magnitude star ε Virginis (Vindemiatrix) lies a string of galaxies within easy grasp of a small telescope on a moonless night. (See the finder chart opposite.)

scope. In fact, I've been able to view all those belonging to Charles Messier's 18th-century deep-sky catalog with 14 × 70 binoculars. Let's focus on a handful we'll find strung in a chain trailing westward from Vindemiatrix, Epsilon (ε) Virginis.

To begin, select an eyepiece that gives your telescope a power of about 50× and place golden Vindemiatrix at the southern edge of the field of view. Then scan 1.6° west, where you'll spot the pretty double star **Σ1689**. This very wide pair, easily split at low power, consists of a 7th-magnitude primary guarding a 9th-magnitude secondary to the southwest. Their spectral classes of *M*4 and *F*5 indicate colors of orange and yellow-white, but they seem golden and bluish to me.

Σ1689 hovers over the roof of a house outlined by 9th- and 10th-magnitude stars. Placing the double at the eastern edge of your 50× field will allow you to encompass the entire domicile. The star marking the pointy peak of the house is due west of the double. From there, the stars making the attic floor lie northwest and southwest. The base of the house rests 1/3° southwest of this, but the southerly corner is deformed by the presence of two widely spaced stars.

The lovely edge-on spiral galaxy **NGC 4762** lies right between those two stars. With my 4.1-inch refractor at 127×, I see a bright needle of light quite elongated north-northeast to south-southwest and surrounded by a thin, fainter halo. The galaxy has a very small, bright core with a starlike nucleus. An 11th-magnitude star sits 2′ south of center.

Another spiral lies west-northwest in the same high-power field — under the house's floor. **NGC 4754** looks almost as bright as its neighbor but much smaller. It's an oval ball of fuzz with a weak central brightening and stellar nucleus. Two faint field stars to the southwest point at the galaxy. Massachusetts amateur Lew Gramer calls this pretty galaxy pair the Spindle and Wool.

Now, putting this pair in the southern part of your 50× field, sweep 2.4° west to the Messier galaxy pair M59 and M60. In my little refractor at 127×, **M60** is the more prominent of the two. It appears round and grows considerably brighter toward the center, where a

Top left: This attractive galaxy duo, NGC 4762 (left) and NGC 4754, are spaced just ¹/₆° apart on the sky (one-third the full Moon's width), so they can be viewed simultaneously in a telescope. Note that five stars in this photograph also appear on the facing finder chart. *Top right:* The two brightest galaxies in this view are M60 (lower left, with companion galaxy NGC 4647) and M59 (upper right). They are ¹/₂° apart, and centered below them is NGC 4638. *Bottom left:* Identifying the galaxy M58 is a snap, thanks to an 8th-magnitude star in the same high-power field. Like all photographs in this article, this view is oriented with celestial north up and west toward the right. *Bottom right:* Spiral galaxy M90 lies ¹/₄° northwest of an 8th-magnitude star (lower-left corner). M90's companion smudge, IC 3583, won't be seen in a small scope.

minute nucleus resides. If the sky is very dark, you may notice a tiny, dim smudge just beyond the northwestern edge of M60's halo: **NGC 4647.** (Using averted vision can be a big help; look off to one side of the galaxy's position instead of directly at it.) Although they look similar in shape, NGC 4647 is a face-on spiral, while M60 is an elliptical galaxy.

The elliptical galaxy **M59** also joins the scene in the 37′ field of my 127× wide-angle eyepiece. This one is distinctly oval and contains a very small, bright core. If you succeeded in finding M60's little companion, try looking for **NGC 4638,** a faint lenticular galaxy in the same field.

Placing M59 in the southern part of a 50× field and scanning 1.1° west brings us to **M58.** Low powers will encompass both. At 127× this barred spiral is slightly oval with a distinct core and a small, bright nucleus. An 8th-magnitude star shines 8′ west of center. With his 4-inch refractor in Hawaii, Stephen James O'Meara has been able to detect hints of M58's central bar. M58 is parked at one corner of a fairly large isosceles triangle framed by three Messier galaxies. M87 and M90, both brighter than M58, sit at the other two corners. In a small scope capable of magnifications under 25×, the three can be seen together.

The elliptical galaxy **M87** is the brightest of the three. Through my little refractor at 127× it appears slightly oval and measures about 4′ by 5′. M87 brightens steadily toward the center and harbors a stellar nucleus. A 9th-magnitude star glitters beyond the northern edge. Small, round **NGC 4478,** a faint companion galaxy with a stellar nucleus, is seen 9′ west-southwest of M87's center.

The spiral galaxy **M90** is a bit dimmer than M87 and more elongated. At the same

magnification it looks about 5′ by 2′ aligned north-northeast to south-southwest. M90 shows a faint halo, a brighter oval core, and a small, bright nucleus.

Within this galaxy triangle we can find yet another Messier object. **M89** is the dimmest and smallest member of the quartet as viewed through my little scope. This elliptical galaxy is round with a fairly bright, very tiny nucleus. It fades outward from the center.

Many other galaxies ornament the wings of Virgo with their softly glowing stardust. They abound in such richness that it's easy to become lost among them. A good star chart (see the Resources section on page 161) will help you find your way around this galaxy-filled region.

A Double Star and 11 Galaxies in Virgo

Object	Type	Mag.	Size/Sep	RA	Dec.
Σ1689	Double star	7.1, 9.1	29″	12ʰ 55.5ᵐ	+11° 30′
NGC 4762	Barred spiral	10.3	8.7′ × 1.6′	12ʰ 52.9ᵐ	+11° 14′
NGC 4754	Barred spiral	10.6	4.4′ × 2.5′	12ʰ 52.3ᵐ	+11° 19′
M60	Elliptical	8.8	7.2′ × 6.2′	12ʰ 43.7ᵐ	+11° 33′
NGC 4647	Spiral	11.3	2.9′ × 2.3′	12ʰ 43.5ᵐ	+11° 35′
M59	Elliptical	9.6	5.1′ × 3.4′	12ʰ 42.0ᵐ	+11° 39′
NGC 4638	Lenticular	11.2	2.5′ × 1.7′	12ʰ 42.8ᵐ	+11° 27′
M58	Spiral	9.7	5.4′ × 4.4′	12ʰ 37.7ᵐ	+11° 49′
M87	Elliptical	8.6	7.2′ × 6.8′	12ʰ 30.8ᵐ	+12° 23′
NGC 4478	Elliptical	11.5	1.7′ × 1.4′	12ʰ 30.3ᵐ	+12° 20′
M90	Spiral	9.5	9.5′ × 4.7′	12ʰ 36.8ᵐ	+13° 10′
M89	Elliptical	9.8	5.0′ × 4.6′	12ʰ 35.7ᵐ	+12° 33′

The galaxies are all members of the Virgo Cluster, which lies about 60 million light-years from our own Milky Way galaxy. They usually appear smaller than the catalog sizes listed.

Markarian's Chain

STRUNG ALONG a 1.5° arc that straddles the Virgo–Coma Berenices border are eight galaxies known as Markarian's Chain. Two of its members (M84 and M86) were spotted in 1781 by Charles Messier, while the rest are best known by their numbers from Johann L. E. Dreyer's 1888 *New General Catalogue* (NGC 4435, 4438, 4458, 4461, 4473, and 4477). But the moniker for the whole group arises from a paper titled "Physical Chain of Galaxies in the Virgo Cluster and Its Dynamic Instability" by Russian astrophysicist Benjamin E. Markarian (*Astronomical Journal:* December 1961, page 555).

I think my most appealing view of the area is through a 10-inch (254-millimeter) reflector with an eyepiece that gives me 43× and an 89′ field. I can capture all of Markarian's Chain plus five or six other galaxies in a single look. Let's visit each in turn. Although a low-power view with its wealth of galaxies is captivating, my descriptions were made at moderate powers of 75× to 100× to help tease out more detail. These galaxies are also visible in my 4.1-inch refractor, but some are challenging to pick out.

Two galaxies known as the Eyes lie practically in the middle of Markarian's Chain. NGC 4435 sits 5′ north-northwest of larger NGC 4438, which shows distortions in its symmetry due to the gravitational interaction between the pair.

Our first target will be **M84**, which sits at the western end of Markarian's Chain, halfway between Vindemiatrix (ε Virginis) and Denebola (β Leonis). My 4.1-inch scope shows a slightly oval glow, aligned northwest to southeast, enveloping an intense round core. The 10-inch reflector exposes a tiny bright nucleus and reveals more of the halo before it fades into the background sky.

M86 lies just 17′ to the east. With my little refractor, it appears larger and slightly dimmer than M84. Aligned in the same direction as its companion, this more elongated galaxy brightens considerably toward its center. The view through the 10-inch adds a stellar nucleus.

Several galaxies that are not part of Markarian's Chain lie in the immediate area. **NGC 4388** is south of M84 and M86 and forms an equilateral triangle with them. This misty east–west slash is fairly easy in the refractor, but it's nearly featureless and shows only a subtly brighter, extended core. The 10-inch picks out a slight bulge in the core, offset to the west of center, and a faint stellar nucleus.

NGC 4387 marks the middle of the triangle formed by the above galaxies. Through the 4.1-inch it is a minuscule bit of fuzz, but the 10-inch turns it into a small oval with a brighter heart. Together these galaxies make up what I think of as the Great Galactic Face. M84 and M86 are its eyes, NGC 4388 forms the mouth, and NGC 4387 indicates the nose. Our face even seems to have one raised eyebrow: **NGC 4402,** a highly elongated spindle similar to NGC 4388 but considerably dimmer. It is a very difficult catch for the little refractor. The larger reflector shows a nearly uniform glow with barely detectable hints of mottling. This eyebrow gives our face a quizzical look, as though its owner is wondering who shaved off the other one.

Two additional galaxies in the vicinity could be mistaken for faint stars in the 4.1-inch. **NGC 4413** is the more difficult of the pair. East of NGC 4388, look for a 10.9-magnitude star with an 11.5-magnitude star 1.6′ to its south-southeast. NGC 4413 is centered due south of the second star by about the same distance and looks like a dimmer fuzzy star. **NGC 4425** is a little easier. Look for it 4′ west-southwest of the brightest star just east of the face.

Returning to members of Markarian's Chain, we'll wander eastward from M86 to **NGC 4435** and **NGC 4438**. In February 1955 *Sky & Telescope* carried an article called "Adventuring in the Virgo Cloud" by Leland S. Copeland, who had been that magazine's first Deep-Sky Wonders columnist in the 1940s. Copeland made a chart, which he called Coma-Virgo Land, and labeled it with fanciful names for patterns of stars and galaxies. He dubbed this pair of galaxies the Eyes, and the name stuck.

Both are easy targets for the small refractor. NGC 4435 is a small oval aligned almost north–south, and it harbors a bright stellar nucleus. NGC 4438 is larger, more highly elongated, and tipped more to the north-northeast. A small bright core lies at its heart. The 10-inch increases the apparent size of the galaxies and adds a faint stellar nucleus to NGC 4438.

Recent observations with the Chandra X-ray Observatory indicate that this galaxy pair underwent a high-speed collision about 100 million years before our current view of them. This distorted NGC 4438 and ejected much of the hot gas that formerly belonged to NGC 4435. NGC 4438 may also have an active galactic nucleus, wreaking further havoc.

If you move 21′ east-northeast of this pair, you'll find another galactic duo, **NGC 4458** and **NGC 4461.** With the 4.1-inch, NGC 4458 appears small, round, and quite faint. NGC 4461 is brighter and elongated nearly north–south. The 10-inch makes NGC 4458 easier to spot and shows a brightish core within a faint halo. NGC 4461 displays a stellar nucleus within a small, round core.

Crossing the constellation border from Virgo into Coma Berenices, we next come to **NGC 4473.** A very easy target for the 4.1-inch, this galaxy is an east–west oval of mist with a starlike nucleus. The 10-inch increases the apparent size of the galaxy and shows that it fades gradually outward from the center.

The final galaxy in Markarian's Chain is **NGC 4477.** It is another easy capture for the small refractor, which shows it to be small and round with a brighter center. Both the fringe and the core of the galaxy look slightly oval in the 10-inch, and a stellar nucleus sits within.

To fit both NGC 4477 and M84 in the same low-power field of the 10-inch, I must put them on opposite edges of the field with part of their halos cut off. With some careful maneuvering I can also lay **NGC 4479** along the edge. But this is a faint galaxy, hard to pick out at the edge of a low-power field.

The reality of Markarian's Chain as a true physical system is still debatable, but that surely won't keep skygazers from enjoying this wonderfully rich area of the sky.

Galaxies of Markarian's Chain

Galaxy	Mag.	RA	Dec.
M84	9.1	12ʰ 25.1ᵐ	+12° 53′
NGC 4387	12.1	12ʰ 25.7ᵐ	+12° 49′
NGC 4388	11.0	12ʰ 25.8ᵐ	+12° 40′
NGC 4402	11.7	12ʰ 26.1ᵐ	+13° 07′
M86	8.9	12ʰ 26.2ᵐ	+12° 57′
NGC 4413	12.3	12ʰ 26.5ᵐ	+12° 37′
NGC 4425	11.8	12ʰ 27.2ᵐ	+12° 44′
NGC 4435	10.8	12ʰ 27.7ᵐ	+13° 05′
NGC 4438	10.2	12ʰ 27.8ᵐ	+13° 01′
NGC 4458	12.1	12ʰ 29.0ᵐ	+13° 15′
NGC 4461	11.2	12ʰ 29.1ᵐ	+13° 11′
NGC 4473	10.2	12ʰ 29.8ᵐ	+13° 26′
NGC 4477	10.4	12ʰ 30.0ᵐ	+13° 38′
NGC 4479	12.4	12ʰ 30.3ᵐ	+13° 35′

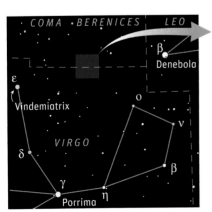

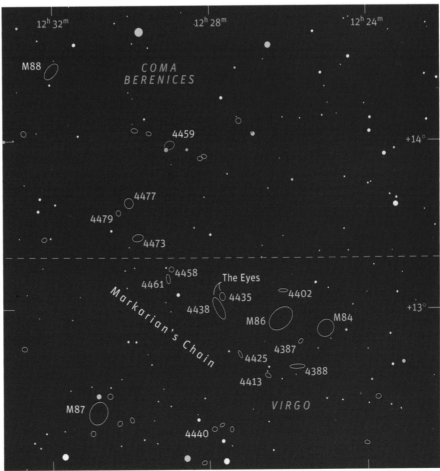

Dazzling Doubles, Glittering Globulars

THE WESTERN HALF of the constellation Serpens (Serpens Caput) winds its way across the early-summer skies between Ophiuchus (aply called the Serpent Bearer) and Boötes, the Herdsman. Hidden among the stars of the Serpent is one of the finest — but often overlooked — globular clusters in our sky. Let's begin our tour with this isolated object. From there we'll hop along a bridge of double stars to a close pair of globulars in nearby Ophiuchus.

M5 is the brightest globular cluster north of the celestial equator. It is located 7.7° southwest of Serpens's brightest star, Alpha (α) Serpentis. At magnitude 5.8, it is easily visible in a 2.4-inch (60-millimeter) telescope as a small, round glow with a fainter, granular halo. A 5th-magnitude yellow star, 5 Serpentis, lies just 0.3° to its south-southeast in the same field of view. A 4-inch telescope at 150× resolves many stars laced around the cluster's fringes. The core appears as a bright blaze intensifying toward a starlike nucleus. The cluster looks slightly elliptical, with the long dimension oriented north-

Double-star observer Sissy Haas sketched 5 Serpentis (Σ1930), Σ1985, and Σ2031. These three figures suggest how the stars appeared in her small refractor. North is up.

east-southwest.

M5 is 13 billion years old — one of the most ancient globular clusters known. It is also among the largest at 130 light-years across. It was first spotted in 1702 by the German astronomer Gottfried Kirche. M5 is a gravitationally bound horde of hundreds of thousands of stars, about 24,000 light-years away.

One of M5's many variable stars can be found in the cluster's halo 3′ southwest of the nucleus. It may be a challenge to find, but it's worth the hunt. **V42** pulses in brightness from visual magnitude 10.6 to 12.1 and back over a period of 25.7 days. At maximum it is the brightest star in the halo, while at minimum it may fade from view through a small telescope.

5 Serpentis, that bright star to the south-southeast of M5, is actually a double. The 5th-magnitude primary is a sun-yellow star. Its 10th-magnitude companion lies 11″ to the northeast and has an orange hue that may be difficult to discern through a small telescope. The two are cleanly split at 80×. This double star is also known as Σ1930. The Greek letter Σ (Sigma), which is found in many double-star designations, indicates that the object is listed in the *Micrometric Measurement of Double and Multiple Stars* catalog compiled by double-star pioneer Friedrich Georg Wilhelm Struve and published in 1837.

Another yellowish pair lies just 2° northeast of the 3.5-magnitude star Mu (μ) Serpentis. **Σ1985** features a 7th-magnitude primary with an 8th-magnitude secondary 6″ to the north. It takes around 120× to comfortably split the two.

About 5° east of Σ1985 we find a third Struve double, near the naked-eye stars Delta (δ) and Epsilon (ε) Ophiuchi, also known as Yed Prior and Yed Posterior. *Yed* comes from the Arabic word for *hand*; these stars mark the left hand of Ophiuchus as he grasps the Serpent. Yed Prior is the leading (western) star of the hand, while Yed Posterior follows it in the stars' unceasing march across the sky. To find **Σ2031,** look 2.1° north-northeast of Yed Prior. This wide double is easily

A pair of globular clusters lie at the heart of Ophiuchus. In this image by Akira Fujii, north is up, M12 is at the upper right and M10 is to the lower left.

separated at 50×, but it may prove challenging since the companion is very faint and nearly colorless through a small telescope. Look for it 19″ to the southwest of the 7th-magnitude yellow primary.

Our Struve steppingstones to the globulars of Ophiuchus have swept us progressively farther from home. Σ1930, Σ1985, and Σ2031 lie at 81, 123, and 152 light-years, respectively. Our jump to the globular cluster **M12** takes us about 16,000 light-years. M12 is located 7.7° east of Σ2031. It is visible through binoculars or a finder as a tiny patch of fuzz. A 4-inch scope at high powers can resolve some of the stars in the outer halo, while the core remains a mottled patch of mist. At 200× a 6-inch scope can reveal several dozen stars, with many in the halo arranged in meandering chains framing dark, starless voids.

Just 3.3° southeast of M12 we find a similar cluster, **M10,** which may be easier to locate than its companion because it is only 1° west of the orange 4.8-magnitude 30 Ophiuchi. If you have trouble sweeping up M12, look for M10 first and proceed from there. The twosome can be seen together through a finder.

M10 appears slightly brighter than its companion and a little more concentrated toward its center. Through a 4-inch scope at high powers the center looks granular, and some stars are resolved around the edges. A 6-inch can show dozens of stars strewn across a hazy background. M10 is 14,000 light-years away, making our two Ophiuchus globu-

lars relatively close neighbors. If the distances are correct, each would be about a 2th-magnitude object in the other's sky.

In the Realm of the Serpent and the Serpent Holder

Object	Type	Mag.	Dist. (l-y)	RA	Dec.
M5	Globular cluster	5.8	24,000	15^h 18.6^m	+2° 05′
V42	Variable star	10.6–12.1	24,000	15^h 18.6^m	+2° 05′
5 Ser (Σ1930)	Double star	5.0, 10	81	15^h 19.3^m	+1° 45′
Σ1985	Double star	7.0, 8.1	123	15^h 56.0^m	−2° 10′
Σ2031	Double star	7.2, 11	152	16^h 16.3^m	−1° 39′
M12	Globular cluster	6.6	16,000	16^h 47.2^m	−1° 57′
M10	Globular cluster	6.6	14,000	16^h 57.1^m	−4° 06′

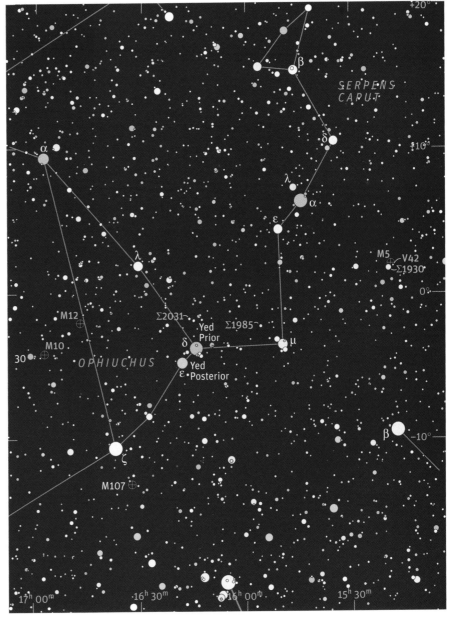

The glittering globular cluster M5 rides the serpent's back in solitary splendor. Less than ½° to its southeast is the double star 5 Serpentis. The variable star V42 lies 3′ southwest of the cluster's center. North is up.

Polar Nights

The sad and solemn night
Hath yet her multitude of cheerful fires;
The glorious host of light
Walk the dark hemisphere till she retires. . . .
And thou dost see them rise,
Star of the Pole! and thou dost see them set.
Alone, in thy cold skies,
Thou keep'st thy old unmoving station yet.

— WILLIAM CULLEN BRYANT
Hymn to the North Star

DIM THOUGH IT MAY BE at 2nd magnitude, the North Star is the most famous in the sky. **Polaris** stands essentially still above your landscape all night and all year (assuming you're in the Northern Hemisphere), holding its station while all others revolve around it. Polaris marks not only true north (to better than ³/₄°) but also your latitude on Earth — which nearly equals Polaris's altitude above your horizon.

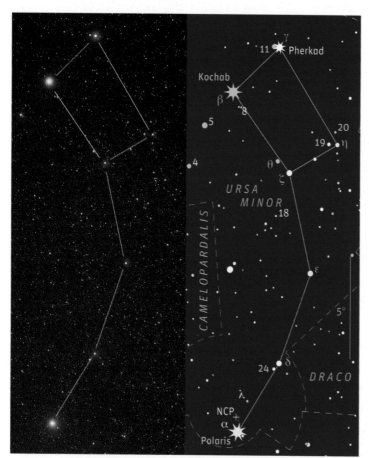

The Little Dipper extends from Polaris to the Guardians of the Pole, Beta (β) and Gamma (γ) Ursae Minoris — a span of nearly 20°. All illustrations are oriented with north down, to match how the Little Dipper stands on late-spring evenings.

Observationally, Polaris has more to offer us than marking the way north. A small telescope at low power will show that it's part of a 40′ circlet of mostly 6th- through 9th-magnitude stars extending in the direction toward Perseus. This asterism is sometimes called the **Engagement Ring,** with Polaris as its sparkling diamond. The Engagement Ring seems to have suffered some rough treatment, however, being oval and badly dented on its southern side.

Polaris itself is a double star; the 2nd-magnitude primary has a 9th-magnitude companion spark a generous 18″ away. It seems our abused Engagement Ring also has a chip off its diamond. This would be an easy low-power split if the stars were more alike, but their difference of 7 magnitudes (a difference of about 600 times in brightness) means the companion is easily overpowered by the brighter star's glare. Try a magnification of at least 80× to bring it out. Polaris itself is pale yellowish. The faint companion looks pale blue by contrast, but measurements show that their colors are actually very similar.

Polaris enjoys its special position because the Earth's axis happens to point toward it. In other words, it stands almost directly above the North Pole. As the world turns, the geographic poles are the only places that face the same direction all night long. Consequently, the star straight above the pole is the only one that seems to stand still. Yet Polaris is not quite the un-moving cynosure described by poets. The axis of our planet misses it slightly; the true north celestial pole is currently about ³/₄° away, in the direction from Polaris toward the Little Dipper's other 2nd-magnitude star, **Kochab** or **Beta (β) Ursae Minoris,** Our North Star travels in a little circle around the true pole each day.

The circle will shrink to just under ¹/₂° radius in the coming century. Why? The Sun and Moon tug on the equatorial bulge of the tipped Earth and try to pull it toward the ecliptic plane. The resulting torque on the spinning planet causes *precession,* a slow change in the orientation of the Earth's axis with respect to the stars. Precession carries the celestial pole around the sky in a large circle that takes some 26,000 years to complete.

So when will Polaris be closest to the north celestial pole? The answer isn't simple. Variations in the gravitational effects of the Moon add a tiny, 18.6-year nodding motion (nutation) to the celestial poles as they tour the precessional circle. Superimposed on nutation is a tiny annual shift in the apparent positions of all stars due to Earth's 30-kilometer-per-second orbital motion with respect to their incoming light (annual aberration). And Polaris changes its own position (proper motion) because of its sideways motion with respect to the Sun in space.

Taking all this into account, computational wizard

Jean Meeus finds that Polaris will be closest to the north celestial pole on March 24, 2100, with an apparent separation of 27′ 09″. While this is probably not the final word on Polaris as the pole star, it shows how a simple question has an ever more complicated answer depending on how precise you want the answer to be.

Among the other stars making up the Little Dipper are four very wide pairs. None is a true binary, but two of them offer attractive color-contrast duos.

One pair is formed by **Zeta (ζ)** and **Theta (θ) Ursae Minoris,** where the Little Dipper's handle joins its bowl. Fourth-magnitude Zeta is white; 5th-magnitude Theta is orange. Separated by 49′, they will fit together only in a low-power, wide-angle view.

In the opposite corner of the Little Dipper's bowl are 3rd-magnitude Pherkad and 5th-magnitude Pherkad Minor, **Gamma (γ)** and **11 Ursae Minoris,** respectively. They're 17′ apart and have hues similar to our previous pair, but the dimmer star is very slightly paler orange, with a color index of 1.4 compared to Theta's 1.6. (A color index of +0.2 is pure white; lower and negative values are slightly bluish, while larger values denote yellow, orange, and red.) Use your lowest power to keep each pair as close together as possible to heighten apparent color contrasts.

The corner of the bowl opposite bright Kochab is formed by 5th-magnitude **Eta (η)** and **19 Ursae Minoris.** They're 26′ apart and tinted, respectively, white with the barest hint of yellow and white with a trace of blue. Can you detect the difference? Forming an isosceles triangle with them is 6th-magnitude **20 Ursae Minoris,** which is yellow-orange.

The fourth pair is **Delta (δ)** and **24 UMi,** the first stars down the handle from Polaris. They are magnitudes 4 and 6, separated by 23′, and both white.

North of the Little Dipper's bowl is a pretty asterism discovered by Pennsylvania amateur Tom Whiting. He dubbed it the **Mini-Coathanger** for its likeness to the familiar Coathanger pattern in Vulpecula (see page 126). It is

composed of ten 9th- through 11th-magnitude stars 1.9° south-southwest of **Epsilon (ε) Ursae Minoris.** Like its larger cousin, the Mini-Coathanger resembles an old-fashioned coathanger with a straight wooden bar and metal hook. Seven stars make up the bar, which is about 17′ long. Three faint stars comprise the hook.

Stars in the Little Dipper

Star	Mag.	Spectral type	Color index	Dist. (l-y)	RA	Dec.
Polaris A	1.9–2.1	F5 – 8 Ib	+0.6	430	2ʰ 31.8ᵐ	+89° 16′
Polaris B	9.0	F3 V	+0.4	430	2ʰ 31.8ᵐ	+89° 16′
δ UMi	4.4	A1 V	0.0	180	17ʰ 32.2ᵐ	+86° 35′
24 UMi	5.8	A2	+0.2	155	17ʰ 30.8ᵐ	+86° 58′
ε UMi	4.2	G5 III	+0.9	350	16ʰ 46.0ᵐ	+82° 02′
ζ UMi	4.3	A3 V	0.0	375	15ʰ 44.1ᵐ	+77° 48′
θ UMi	5.0	K5 III	+1.6	800	15ʰ 31.6ᵐ	+77° 21′
η UMi	5.0	F5 V	+0.4	97	16ʰ 17.5ᵐ	+75° 45′
19 UMi	5.5	B8 V	−0.1	660	16ʰ 10.8ᵐ	+75° 53′
20 UMi	6.4	K2 IV	+1.2	760	16ʰ 12.5ᵐ	+75° 13′
γ UMi	3.0	A3 III	+0.1	480	15ʰ 20.7ᵐ	+71° 50′
11 UMi	5.0	K4 III	+1.4	390	15ʰ 17.1ᵐ	+71° 49′
Kochab	2.0	K4 III	+1.5	125	14ʰ 50.7ᵐ	+74° 09′

Above: How many observers know the faint Engagement Ring asterism that holds Polaris as its jewel? The ring is ²/₃° across, so use your telescope's lowest-power, widest-field eyepiece.

Left: The Mini-Coathanger is a ¹/₃°-long asterism floating near Epsilon (ε) Ursae Minoris between the Little Dipper's handle and bowl.

Grabbing the Bear by the Tail

ON THE STAR CHART on page 25, you'll see the kite-shaped figure of Boötes, the Herdsman, high in the south. His upraised arm reaches out to grab the tail of Ursa Major, the Great Bear — or perhaps Boötes has just let her go, and we see the Bear with her tail stretched long from her struggle to get away. Let's join Boötes in taking the Bear by the tail and see what riches this area of the sky has to offer.

We'll start with the star at the bend of the Bear's tail. **Mizar,** or Zeta (ζ) Ursae Majoris, is surely the sky's most famous naked-eye double. Anyone with reasonably good eyesight should be able to spot 4th-magnitude Alcor close by to the east-northeast. Mizar and Alcor easily fit together in a telescope's low-power field, along with an 8th-magnitude star that makes a squat isosceles triangle with them. In 1722 an eccentric German professor, Johann Liebknecht, mistook this faint star for a new planet and named it Sidus Ludoviciana ("Ludwig's Star") in honor of his king.

Although Mizar and Alcor share a similar distance and motion across the sky, they probably do not form an orbiting pair. Instead, it seems more likely that they are slowly separating members of a star cluster known as Collinder 285, the Ursa Major moving cluster. The central five stars of the Big Dipper all belong to this widely spread group.

Mizar, however, is a true binary whose nature is revealed at 40× through a small telescope. The 2nd-magnitude primary has a 4th-magnitude secondary 14″ to the south-southeast. Both components are white.

Long-exposure photography transforms M101, a large hazy patch in small telescopes, into one of the most beautiful of all spiral galaxies. The field here is nearly 1° across, with north up.

NGC 5866 appears as an edge-on galaxy in photographs, with a narrow dust lane running through the middle. Less than ¼° to the west-southwest is an 8th-magnitude star (at far right).

Mizar was the first binary star discovered, the first photographed, and the first star found to have an additional companion revealed in details of its spectrum. In fact, *both* of the close visible components of the Mizar system have unseen spectroscopic companions. A fascinating account of Mizar's history can be found on Leos Ondra's Web page at http://leo.astronomy.cz/mizar/article.htm.

Our next target is the galaxy **M101.** It is sometimes called the Pinwheel Galaxy, a name it shares (confusingly) with two other Messier objects: M33 in Triangulum and M99 in Coma Berenices. This face-on spiral is nearly the diameter of the full Moon, but its very low surface brightness makes it a difficult find.

If you have a unit-power (1×) finder on your telescope, note that M101 forms a nearly equilateral triangle with Mizar and the tail-end star, Alkaid or Eta (η) Ursae Majoris. Try to position the finder's red dot or bull's-eye target accordingly. If you have an equatorial mount, you can put Mizar in the northern half of a low-power field and sweep 5.7° eastward. Through a finder, you can follow the line of four 5th- and 6th-magnitude stars that starts east of Mizar and trends east-southeast. Each of these methods should get you close enough to scan for M101 with a low-power eyepiece. Look for an extremely dim, extended glow.

With 14 × 70 binoculars, M101 appears large, faint, round, and nearly uniform with a hint of brightening in the center. With a 3.6-inch (92-millimeter) refractor at 50×, I see a large, roundish glow about 15′ across with a slightly brighter core. A similar power with my 4.1-inch refractor makes the halo look oval. At higher magnification, the core itself appears slightly oval with a tiny nucleus and a faint star near its northern edge. Observers at dark-sky sites may see some patchiness in the halo.

The hand that's grabbing for the Bear's tail contains the interesting double stars **Kappa (κ)** and **Iota (ι) Bootis**. Kappa consists of a 4.5-magnitude primary with a 6.6-magnitude secondary 14″ to the southwest. The pair are nicely separated at 47×. The brighter star has a spectral class of *A*7, indicating that it is a white star. The companion's spectral class is *F*1, meaning it is white with a slight tinge of yellow. Despite this, some observers see the secondary as bluish. It seems very common for stargazers to perceive a star as bluish when it lies near a brighter white, yellow, or orange star. What color do you see?

Iota's components have a greater color difference. The 4.8-magnitude primary has a 7.4-magnitude secondary 40″ to the northeast, easily split at low power. Their spectral classes are *A*7 and *K*0, respectively, implying colors of white and yellow-orange. Nonetheless, most observers report seeing the brighter star as yellow and the dimmer as blue or white. How about you?

Now we'll visit M102 — or perhaps we already did. Pierre Méchain supposedly discovered M102 in 1781, and Charles Messier included it as a last-minute addition to his famous catalog even though he'd not had time to observe it yet. In 1783 Méchain disowned his discovery in a letter to Johann Bernoulli in Berlin. He wrote: ". . . this is nothing but an error. This nebula is the same as the preceding No. 101. In the list of my nebulous stars communicated to him, M. Messier was confused due to an error in the sky chart."

This would seem to lay the existence of M102 to rest, but some still find the point debatable. Made two years after the fact, could it be that Méchain's retraction was itself in error? German astronomer Hartmut Frommert makes a reasonable case for this possibility. Following a trail of evidence that includes Messier's personal notes on a later observation of M102, descriptions by both Messier and Méchain, and some speculative detective work, Frommert proposes NGC 5866 as the most likely candidate for M102 — if it was not just a mistake. The full story can be seen at http://www.seds.org/messier/m/m102d.html.

To locate **NGC 5866,** look for the naked-eye star Iota (ι) Draconis. The galaxy is exactly 4° southwest of this golden star, close enough to share a finder field. Use the chart at right to help you star-hop to our goal. Through 14 × 70 binoculars the galaxy appears quite small, very faint, and elongated. It lies near one point of a right triangle of nearly equal stars, and a bright star in Boötes can be seen to its south. With my 6-inch reflector, NGC 5866 is about 2′ long and one-third as wide (elongated northwest to southeast),

growing brighter toward the long axis. Two faint stars bracket the galaxy's northwestern end. Messier object or not, NGC 5866 is a delightful sight for the finale of our tour.

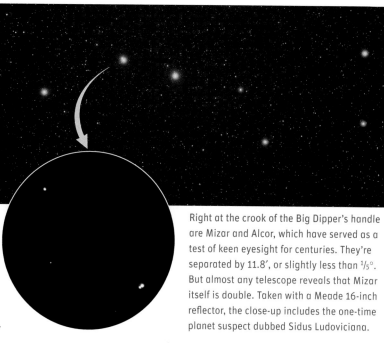

Right at the crook of the Big Dipper's handle are Mizar and Alcor, which have served as a test of keen eyesight for centuries. They're separated by 11.8′, or slightly less than ⅕°. But almost any telescope reveals that Mizar itself is double. Taken with a Meade 16-inch reflector, the close-up includes the one-time planet suspect dubbed Sidus Ludoviciana.

Springtime Sights Between Boötes and the Bear

Object	Type	Mag.	Size/Sep.	Dist. (l-y)	RA	Dec.
Mizar	Double star	2.2, 3.9	14.4″	78	13ʰ 23.9ᵐ	+54° 56′
M101	Galaxy	7.9	27′ × 26′	26 million	14ʰ 03.2ᵐ	+54° 21′
κ Boo	Double star	4.5, 6.6	13.7″	155	14ʰ 13.5ᵐ	+51° 47′
ι Boo	Double star	4.8, 7.4	39.9″	97	14ʰ 16.2ᵐ	+51° 22′
NGC 5866	Galaxy	9.9	6′ × 3′	38 million	15ʰ 06.5ᵐ	+55° 46′

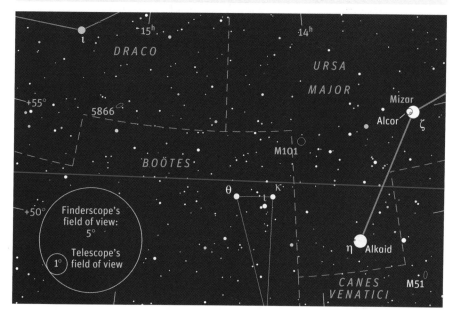

Spectral Hues of Double Stars

DOUBLE STARS are wonderful targets for small telescopes, even in less-than-ideal skies. While getting ready for this essay, I encountered a stubborn spell of cloudiness that offered but one observing opportunity — at full Moon, with twinkly skies at –10° Fahrenheit. Yet the stars were lovely, for I was going after doubles bright enough to reveal various hues under even those conditions.

Color perception is a complicated issue involving physiology, psychology, and observing conditions. It's little wonder that skygazers report different colors for the same stars. But I had a great advantage when I surveyed the stars described here with my 4.1-inch refractor. Since I had picked the stars, I knew what colors to expect. I also find that first impressions are often the best. After staring at a subtly colored star for a while, my eyes become less sensitive to the color.

I started by looking up the spectral types of my targets in the SIMBAD astronomical database. The first letter of a star's spectral type roughly correlates with its surface temperature. From hottest to coolest the main types are O, B, A, F, G, K, and M, and their colors trend from blue-white through white, yellow, orange, and finally red-orange. Most spectral types end in the Roman numeral V, referring to the luminosity class of main-sequence stars, like our Sun, that are fusing hydrogen in their cores. The Roman numerals II and III denote aged giant stars, and a type ending with the small letter p means the spectrum is peculiar.

Next, I checked their color indices in the European Space Agency's *Hipparcos and Tycho Catalogues*. The color index is the difference between a star's apparent magnitudes at two wavelengths. For blue and yellow-green light, the values run from about –0.4 for spectral type O to +1.7 for type M, giving me an independent clue as to the hue. In most cases the two indicators agreed.

The doubles we'll visit are in the constellations Boötes and Corona Borealis, high in the south on the all-sky chart on page 25. Our first will be **Σ1835** in the far-southern reaches of Boötes. Easily swept up in a finderscope, it is the northernmost of a little arc of three 5th- and 6th-magnitude stars lying about 11° south of Arcturus. Observers have described the primary star as bluish, greenish, white, or pale yellow, and the secondary as various shades of blue, white, yellow, or even lilac. To me they simply appeared white and yellow when split at 87×. I found that a slight defocusing made the stars' colors more apparent.

Now we'll move north-northeast to the naked-eye star **Pi (π) Boo.** Stargazers agree fairly well on the color of the brighter star, citing shades of blue-white or white, but opinions about the companion cover nearly the entire range of possibilities. At 87×, I see these stars as blue-white and yellow-white. The dimmer star is truly more of a pure white, but contrast with the bluer primary may make it seem more yellow than it really is.

Xi (ξ) Boo lies northeast of Pi and is also visible to the unaided eye. The brighter component has been seen clothed in shades of yellow or orange, and though the attendant has been logged similarly, several observers have noted an unusual red-purple cast. To me at 87×, they appear orange and red-orange.

Next we'll visit **Σ1825,** visible in a finder just 1° north of Arcturus. Its primary has been depicted with tints of pale blue, yellow, or white, and the companion with bluish, white, or orange. At 127×, I saw the bright star as white. I couldn't turn up a spectral type for the secondary, but its color index indicates a yellow star. At times it did appear yellow, but through the turbulent atmosphere it sometimes twinkled with a bluish light or a colorless gray. Dim stars are often mistakenly perceived too blue.

Epsilon (ε) Boo, or Izar, is the most brilliant of our doubles, and it has justly earned its nickname Pulcherrima (Latin for "most beautiful"). The primary is generally seen in shades of yellow and orange. The secondary has been called many different colors, most commonly blue, green, or a blend of the two — but occasionally purple or reddish purple! Among the stars described here, this was the most difficult for me to split, mainly because of its brightness. At 127×, atmospheric turbulence turned the stars into overlapping blobs of orange and white. But Izar is one of my favorite doubles, and I've had pleasing views at 87× when the air is steady.

Mu (μ) Boo makes a splendid pair, and the star known as μ¹ is visible to the unaided eye. The colors of μ¹ involve tints of blue, green, white, or yellow, while those of μ² include green, white, yellow, and orange. At 28×, the pair was very wide and looked white and yellow to me. Zooming in on μ² at 127×, I could split it into two very close yellow components.

Hopping 3° east-southeast into Corona Borealis, we come to naked-eye **Zeta (ζ) CrB.** Both members are usually described with tints involving blue, green, or white. At 87×, I see the primary as blue-white and its attendant as very blue. In truth, the stars have almost the same color, but comparison with the brighter star made the dimmer one looker bluer to me.

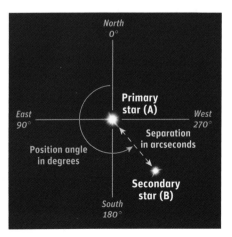

Our next star is **Sigma (σ) CrB.** To locate it, aim your finder at 4th-magnitude Theta (θ) CrB and star-hop east-northeast through a chain of 5th-magnitude stars composed of Pi (π), Rho (ρ), and Sigma. The primary of σ CrB has been called bluish white, creamy white, pale yellow, or yellow. Oddly, the supposedly yellowish secondary has been described as purple, bluish white, and blue. To me, both stars looked yellow at 87×.

Our final target is **Nu (ν) CrB,** a short jump 1⅔° east of Sigma. With fairly bright orange stars, people have less trouble agreeing on color. Descriptions of both ν¹ and ν² include yellow, deep yellow, golden, and orange. At 17×, this wide pair appears red-orange and yellow-orange to my eyes.

Apart from colors, double stars are described by their separation in arcseconds and the *position angle* (p.a.) of the dimmer star with respect to the brighter star. A p.a. of 0° means the companion lies to the north; 90° is east, 180° is south, 270° is west, and so on.

Double Stars to Test Your Color Sense

Object	Mag.	Sep.	P.A.	RA	Dec.	Spectral types		Color indices	
Σ1835	5.0, 6.8	6.2″	194°	$14^h\ 23.4^m$	+08° 27′	A0V	F0V + F2V	0.0	+0.4
π Boo	4.9, 5.8	5.6″	109°	$14^h\ 40.7^m$	+16° 25′	B9p	A6V	0.0	+0.2
ξ Boo	4.8, 7.0	6.4″	315°	$14^h\ 51.4^m$	+19° 06′	G8V	K4V	+0.7	+1.2
Σ1825	6.5, 8.4	4.4″	156°	$14^h\ 16.5^m$	+20° 07′	F8V	—	+0.5	+0.8
ε Boo	2.6, 4.8	2.9″	346°	$14^h\ 45.0^m$	+27° 04′	K0II-III	A2V	+1.0	+0.2
μ¹, μ² Boo	4.3, 7.1	109″	171°	$15^h\ 24.5^m$	+37° 23′	F0V	G1V	+0.3	+0.6
ζ CrB	5.0, 5.9	6.1″	305°	$15^h\ 39.4^m$	+36° 38′	B7V	B9V	−0.1	−0.1
σ CrB	5.6, 6.5	7.1″	237°	$16^h\ 14.7^m$	+33° 52′	G0V	G1V	+0.6	+0.6
ν¹, ν² CrB	5.4, 5.6	361″	164°	$16^h\ 22.4^m$	+33° 48′	M2III	K5III	+1.6	+1.5

The star colors shown in this article's small circular plots are those expected from the spectral types and color indices listed here. Most of the circles are 30″ across; those for μ Boo and ν CrB are much wider.

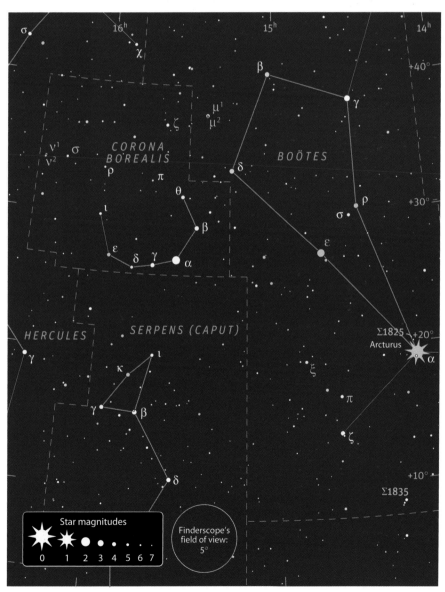

Riding high on June evenings are two constellations renowned for their double stars: Boötes, the Herdsman, and Corona Borealis, the Northern Crown. Star-hop to each star's location with your telescope's finder, which typically shows 5° of sky (as shown). Then switch to a much higher power on your main scope to split the double.

Follow the Arc

BOÖTES, THE HERDSMAN, is an easy constellation to locate. The Big Dipper's familiar pattern leads us to it if we remember the saying, "Follow the arc to Arcturus." Extending the curve of the Big Dipper's handle takes us right to Arcturus, the brightest star in Boötes. The skinny kite-shaped figure of Boötes stretches northeast from this brilliant golden star. One of the Herdsman's arms reaches out toward the handle of the Dipper as if he is clutching the Great Bear by the tail. Perhaps this is how the Great Bear's tail has stretched long as she flees before Boötes in their nightly chase around the Polestar.

Although simple to find, Boötes is a constellation lacking deep-sky wonders — or so you might think. I asked several active observers in my astronomy club if they could think of any deep-sky objects in Boötes, and the only suggestion was Izar, Epsilon (ε) Bootis, the beautiful gold and white double star discussed on page 92. Yet Boötes does hold other sights worth seeking.

We'll start with the **Arcturus Group.** I learned of this gathering of 5th- through 9th-magnitude stars in the fabulous book *Star Clusters* by Brent Archinal and Steven Hynes. The authors unearthed what seems to be the first mention of the group in the 1856 edition of *The Geography of the Heavens* by Hiram Mattison and Elijah H. Burritt. There it is described as "a rich group of stars in the vicinity of Arcturus, and surrounding that star. May be seen with small telescopes." The figure from the companion atlas shows 48 stars, but neither scale nor directions are indicated. Binoculars reveal a roughly corresponding collection of stars. There is a

core group about 3.5° by 2° with the long dimension running east–west, and outliers expand the group to perhaps 5° by 3°. A small telescope will bring out hints of yellow and orange in some of the members, but few scopes can achieve a wide enough field of view to encompass them all. The Arcturus Group is not a true star cluster but rather an asterism of unrelated stars that happen to lie along the same line of sight.

A much smaller asterism lies in the southern edge of the Arcturus Group. **Picot 1** consists of seven stars ranging from magnitude 9.4 to 10.7 distributed along a bell curve. The French amateur astronomer for whom the group is named, Fulbert Picot, gave it the apt nickname Napoleon's Hat. The brim of the hat is about 20′ long and runs northeast to southwest. Almost any telescope can capture the emperor's chapeau.

Moving northward, we come to one of my favorite targets in Boötes, **NGC 5466,** a ghostly globular star cluster with a low surface brightness. Despite this, it can be glimpsed in a dark sky through binoculars. Look for it 6° west-southwest of Rho (ρ) Bootis next to a 7th-magnitude orange star. Through a small scope, NGC 5466 is a weak, round glow about 5′ across. My 4.1-inch (105-millimeter) at 87× can pick out only a few of the cluster's brightest stars, but a 6- to 8-inch scope might catch a dozen. Larger apertures will let you boost the magnification without making the cluster fade away. With my 10-inch reflector at 213×, NGC 5466 is 6′ across and very slightly elongated east–west. More than two dozen very faint stars shimmer over unresolved haze, the brightest strewn mostly around the edges.

One of the most spectacular globular clusters in the northern sky lies 5° due west of NGC 5466. Although a resident of Canes Venatici, not Boötes, **M3** is well worth the side trip. The cluster is bright enough to be seen in binoculars, and some observers have spotted it with the unaided eye. Through my 4.1-inch scope at 127×, M3 appears sparse around the edges with a much brighter core that intensifies toward the brilliant center. Some stars are resolved right down to the cluster's heart. The core, elongated northwest-southeast, is buried in a halo 12′ across. A 10th-magnitude star shines at the northwestern edge, and an 8th-magnitude star lies outside the halo to the south-southeast. This globular takes magnification well, so boost the power to help bring out its fainter stars.

Larger telescopes resolve M3 into a glorious blaze teeming with countless stars. Outliers radiate from the central mass in meandering streams and glittering rays. M3 appears more impressive than its neighbor NGC 5466 because it is intrinsically six times brighter and only two-thirds as far away.

French observer Charles Messier discovered this globular cluster while tracking Comet Bode of 1779, and he made it the third entry (M3) in his famous deep-sky catalog. Yet he saw only a glow here, not any stars. I can resolve individual stars with my modern 4.1-inch refractor. North is up.

Many of the deep-sky denizens of Boötes are galaxies, none terribly bright. One of the most interesting is **NGC 5529,** a remarkably flat galaxy. NGC 5529 is actually a spiral whose thin disk, seen edge on, looks highly elongated or "flat." It is one of the entries in the *Revised Flat Galaxy Catalogue* (Igor D. Karachentsev and colleagues, 1999), a compilation of 4,236 galaxies that appear at least seven times longer than wide. The vast majority of those galaxies are not visible in the average backyard telescope, but NGC 5529 is one exception.

To find it, move 2° west-southwest of Gamma (γ) Bootis to a 7th-magnitude star, the brightest in the area, and then extend your sweep for that distance again. Although NGC 5529 has been seen with instruments as small as 6 inches in aperture, I have never looked at it in anything smaller than my 10-inch Newtonian. At 115× I see its needle-like sliver running west-northwest to east-southeast. The galaxy is a little wider and brighter in the center, and it harbors an extremely faint stellar nucleus. It is flanked by two concentric arcs of three stars each, magnitudes 10 through 12. NGC 5529 is closer to the eastern curve, and its southeastern tip is near the arc's southernmost star.

A much rounder galaxy is found 38′ east-northeast of NGC 5529 and can share the same low-power field. **NGC 5557** has a higher surface brightness and should be easier to see. My 4.1-inch scope at 17× displays a small circular patch that is brighter in the center. At 87× it appears slightly oval with a thin halo surrounding the brighter core. Boosting the magnification to 127× uncovers an elusive, starlike nucleus. Views through my 10-inch enlarge the galaxy a little by revealing more of its halo.

A small group of galaxies lies beside the Herdsman's upraised arm, at least two of them bright enough to be seen in a small telescope. First, look for a 5.7-magnitude reddish orange star 3° south-southeast of Theta (θ) Bootis. From there our first stop will be 19′ west-northwest, where we'll find **NGC 5676.** It is visible through my little refractor at 47× as an extended smudge running northeast to southwest. At 87× it maintains a nearly uniform surface brightness. A 10-inch shows a faint halo three times longer than wide enclosing a large oval core. A brighter patch ornaments the eastern end of the core, a clue that this is a spiral galaxy.

Although slightly fainter, **NGC 5689** is also visible at 47× through the small refractor. It lies 38′ south-southeast of the ruddy star mentioned above and shows an east-west elongation. At 87× I can see that it grows brighter toward the center. A 10-inch at moderate to high power shows that the core is slightly mottled and harbors a stellar nucleus. NGC 5689 is a barred-spiral galaxy viewed nearly edge on.

We can see that Boötes does indeed boast cosmic wonders worth pursuit. So when some star-filled night beckons, be sure to *follow the arc* and herd up a few for yourself.

Buried Booty of Boötes

Object	Type	Size	Mag.	RA	Dec.
Arcturus Group	Asterism	5° or so	—	14ʰ 19.5ᵐ	+19° 04′
Picot 1	Asterism	20′ × 7′	—	14ʰ 14.9ᵐ	+18° 34′
NGC 5466	Globular cluster	9′	9.0	14ʰ 05.5ᵐ	+28° 32′
M3	Globular cluster	18′	6.2	13ʰ 42.2ᵐ	+28° 23′
NGC 5529	Spiral galaxy	6.4′ × 0.7′	11.9	14ʰ 15.6ᵐ	+36° 14′
NGC 5557	Elliptical galaxy	3.6′ × 3.2′	11.0	14ʰ 18.4ᵐ	+36° 30′
NGC 5676	Spiral galaxy	4.0′ × 1.9′	11.2	14ʰ 32.8ᵐ	+49° 27′
NGC 5689	Spiral galaxy	4.0′ × 1.1′	11.9	14ʰ 35.5ᵐ	+48° 45′

Angular sizes are from catalogs or photographs; most objects appear somewhat smaller when a telescope is used visually.

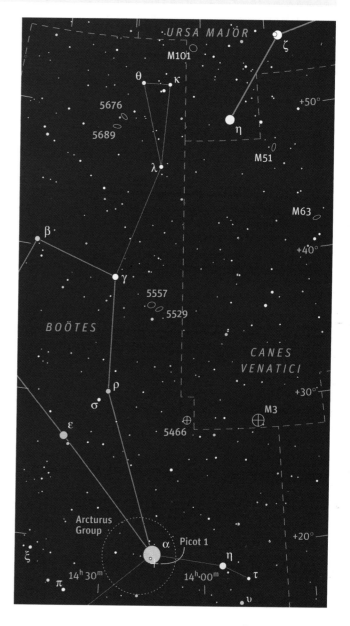

Chapter

The star clouds and nebulae that dwell in the softly glowing band of the Milky Way in Scorpius and Sagittarius are a good place for novice deep-sky hunters to get their feet wet. Elsewhere in the summer sky, many fine globular clusters await inspection. But I'll take you beyond these highlights to some overlooked wonders in seldom-explored constellations like Ophiuchus, Vulpecula, and Draco.

Deep-Sky Wonders in Hercules

THIS MONTH HERCULES appears upside down on our all-sky chart on page 26 with his legs wrapped around the zenith (the map's center) and his head just above Ophiuchus. The central six stars of this celestial strongman form a distinctive butterfly shape — to my eye — that serves as a starting point for some July deep-sky explorations.

Let's begin with Hercules' leading lady, **M13.** Look for it ²/₃ of the way from Zeta (ζ) to Eta (η) Herculis in the butterfly's northern wing. Often called the Great Cluster in Hercules, it was discovered by Edmond Halley, who wrote, "This is but a little Patch, but it shews itself to the naked Eye, when the Sky is serene and the Moon absent."

If your sky is less than serene, you should still be able to see the cluster as a tiny smudge of light through a small finder. M13 may be the easiest globular cluster to resolve into stars through a small telescope from midnorthern latitudes. Through my 4.1-inch (105-millimeter) scope at 127×, M13 is gorgeous! It appears about 15′ across and brightens gradually and considerably toward the center. The central mass contains several hundred thousand stars and resembles a mound of tiny, glittering ice crystals. Some of the outlying stars are arranged in pretty arcs and arms. Al-

The globular cluster M13 is easily located on summer evenings high overhead in the keystone-shaped center of Hercules. The field in this image measures about 42′ across. North is up in all illustrations.

though some northern observers favor the globular M5 (see page 86), M13 rides much higher across the sky, where we can view it through a thinner layer of Earth's image-degrading air. Which do you prefer?

Hercules also harbors many faint galaxies. **NGC 6207** is one of the least faint of the bunch, though it's still a challenge at magnitude 11.6. It lies in the same low-power field as M13, just 28′ northeast of the cluster's core. A 3.6-inch scope at 94× under a dark sky shows it as a small, slightly oval smudge.

M13's globular costar is the often-overlooked **M92.** This lovely cluster lives in the shadow of its famous neighbor, but it's not hard to find once you know where to look. Locate M13 and hop up to the star Eta (η) Herculis. From there, look southeast to Pi (π) Herculis. These two stars are just under 7° apart. Now sweep your gaze north of Pi almost the same distance to find M92 glimmering in the darkness; it's bright enough to be seen in a good finder. Through my 4.1-inch refractor at 127×, M92 appears slightly oval. Its halo is elongated northeast-southwest and is around 8′ × 7′ in size. The core looks round and is about 2.5′ across. Outlying stars are well resolved, and the outer core appears much brighter than the halo. The small, unresolved center is brighter still.

Hercules contains one planetary nebula that might be considered a small-scope tar-

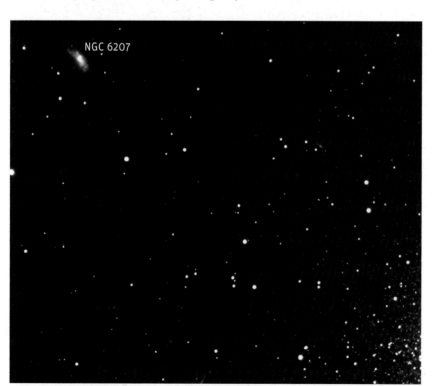

NGC 6207

The faint galaxy NGC 6207 lies about a half degree to the northeast of M13 (just off the lower right in this photo). It looks like an elongated smudge with a bright core.

get. **NGC 6210** lies 4° northeast of Beta (β) Herculis, the southwestern wingtip of the constellation's butterfly shape. Look for it just north of two 7th-magnitude stars in a northeast-southwest line. While NGC 6210 has been seen through 3-inch scopes, it tends to look like a faint, out-of-focus star. In my 6-inch scope at 39×, it's a tiny, robin's-egg-blue disk. At 176× the disk is small, uniform, and oval. The central star of NGC 6210 is elusive and often invisible, even through large telescopes when the air is unsteady. This planetary is sometimes referred to as the Turtle Nebula.

Now let's move to the southeastern wingtip of the butterfly, Delta (δ) Herculis. It marks the western end of a line of naked-eye stars that runs to the east-northeast. Hop from Delta through 4th-magnitude Lambda (λ) to 3rd-magnitude **Mu (μ) Herculis.** Mu is a double star, with a 10th-magnitude secondary 34″ to the west-southwest of the yellow primary. A magnification of 50× will widely split the pair. The secondary is actually composed of two red-dwarf stars too close together to be resolved through small telescopes.

Now hop from Mu through 3.7-magnitude Xi (ξ) and continue for a bit more than that distance again to locate the open star cluster **Dolidze-Džimšelejšvili 9.** With a name like that, how could you pass up a look? Through my 4.1-inch scope at low power, this cluster shows

about 25 stars scattered across ½° of sky. Four of the brightest stars are arranged in two pairs — one in the southern reaches of the cluster, the other in the north. The southern pair shows a nice color contrast, with the brighter star white and the fainter one orange. The cluster's unusual name comes from its discoverers, Madona V. Dolidze and G. N. Džimšelejšvili.

This diverse array of celestial marvels — two globular clusters, a galaxy, a planetary nebula, a double star, and an open cluster — doesn't begin to exhaust the wonders of Hercules! The fifth-largest constellation of the starry vault lies before you with an abundance of dim but enchanting spectacles dwelling in its depths, waiting to be hunted out with your telescope and star atlas.

Herculean Sights

Object	Type	Mag.	Dist. (l-y)	RA	Dec.
M13	Globular cluster	5.8	23,500	16ʰ 41.7ᵐ	+36° 28′
M92	Globular cluster	6.5	25,400	17ʰ 17.1ᵐ	+43° 08′
NGC 6207	Galaxy	11.6	—	16ʰ 43.1ᵐ	+36° 50′
NGC 6210	Planetary nebula	9.3	3,600	16ʰ 44.5ᵐ	+23° 49′
μ Her	Double star	3.4, 10.1	27.4	17ʰ 46.5ᵐ	+27° 43′
Do-Dž 9	Open cluster	—	—	18ʰ 08.8ᵐ	+32° 32′

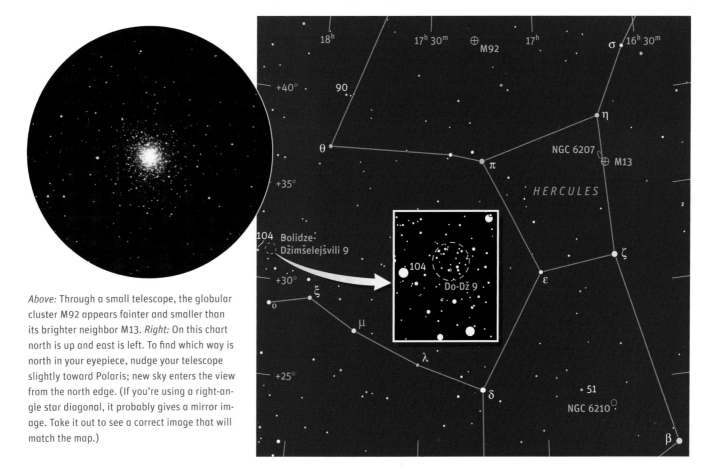

Above: Through a small telescope, the globular cluster M92 appears fainter and smaller than its brighter neighbor M13. *Right:* On this chart north is up and east is left. To find which way is north in your eyepiece, nudge your telescope slightly toward Polaris; new sky enters the view from the north edge. (If you're using a right-angle star diagonal, it probably gives a mirror image. Take it out to see a correct image that will match the map.)

Cuddling Up to the Scorpion

THE CONSTELLATION SCORPIUS appears just above the "Facing South" label on the July all-sky chart on page 26. That means June and July evenings are prime time for having a look. The Scorpion may not be the most lovable creature, but by keeping close to his bright stars we can easily locate a number of interesting sights for small telescopes.

Let's start with the northernmost stars in the stick-figure Scorpion, also shown in the close-up chart at right. **Nu Scorpii** (ν on our charts) is a wide double star for even the smallest of scopes. Like many other doubles, its stars show only a subtle color difference of blue-white and white. But each component *also* has a very close companion, making Nu a quadruple star system. The dimmer star is easier to distinguish as a pair, but you'll probably need a good 4-inch telescope and an eyepiece providing 150× to do so. The brighter one is quite a bit more difficult.

Beta (β) Scorpii, also known as Graffias, is another fine double star, easily split in a small telescope at about 50×. Most observers see both components as either blue-white or white.

Our next targets are two strongly contrasting globular star clusters, M4 and M80. The "M" stands for

Charles Messier, and anything that made this 18th-century observer's list is fair game for newcomers to astronomy. (That's not true for many other listings of interesting sights.) All the telescopes Messier used were quite small; not a single one had as much light grasp as a good 4-inch telescope of today.

The first of these clusters lies next to Antares, the fiery orange heart of the Scorpion. The easiest way to find **M4** is to put a low-power eyepiece in the telescope, center on Antares, and then nudge the tube little by little toward the right. A 4-inch scope at 75× will show many of M4's individual stars, some of which seem to be arranged in a central bar that runs nearly north-south. At 7,000 light-years from Earth, M4 is one of the very closest globular clusters. Its stars seem loosely scattered to me; they aren't as strongly concentrated toward the center as they are in other globulars.

The other cluster, **M80,** is more challenging. Wait for a clear, dark night with no Moon in the sky and set up your telescope as far as possible from city lights. Located halfway between Graffias and Antares, M80 is one of the richest and most compact star clusters in all the heavens. But it lies at such an immense distance, four times as far as M4, that a small telescope can't show its individual stars. At 75× it looks more like a tiny, glowing cloud with a bright core. But you'll have to identify it by location first, because M80 looks just like any 7th-magnitude star in a finderscope.

Wending our way southward along the body of the Scorpion, we come to an impressive group of objects known as the False Comet. Unfortunately, the snowbirds among us need to pack up our small scopes and travel south for a good view of this area. Those at the latitude of Virginia or San Francisco will see it crest in the southern sky a mere 10° above the horizon. The farther south you live, the higher up it gets. In fact, it likely got its nickname back in 1983 at the Texas Star Party, when Alan Whitman commented on "a comet-like Milky Way patch in southern Scorpius."

Because the False Comet spans about 2½° of sky, it is a good example of something easier to appreciate with a small scope or binoculars than with a large instrument. It even has a "head" and a "tail," but they are optical illusions produced by a remarkable collection, or association, of young, hot stars lying about 6,000 light-years away toward the center of our galaxy. They delineate part of the next spiral arm inward from our Sun, the Sagittarius Arm.

The head of the False Comet is formed by the bright double star **Zeta (ζ) Scorpii.** Through a small scope at low power the two stars show a nice color contrast, with the brighter star, called Zeta², appearing orange

By early June each year, Scorpius is rising in the east just as darkness falls in the Northern Hemisphere. Once this constellation has reached due south, its sting is as high above the horizon as it will ever get.

and the dimmer Zeta¹ blue-white. But they aren't at all close together in space. Zeta² is a foreground star and neighbor of ours, for it lies a paltry 150 light-years away. But Zeta¹ is a member of the distant stellar association and must be 100,000 times as luminous as our Sun to appear so bright from so far off.

The tail of the False Comet is formed by two quite different star clusters. The more compact one, known as **NGC 6231** (or Caldwell 76) is an amazing sight. Despite its distance, about 20 of its stars can be seen with a scope of only 2.4-inch aperture. The brightest seven or eight stars have been compared to the Pleiades, but that's somewhat disparaging. If NGC 6231 were brought as close as the Pleiades, each of its bright stars would rival Sirius!

The second cluster, **Trumpler 24,** is much more spread out — about 1½° (three Moon-widths) across. It lies just northeast of NGC 6231 and is best enjoyed when encompassed in a wide-angle field and viewed at a magnification of 20× or less. A 3-inch to 4-inch scope can show about 100 stars here. The brightest are splashed in a wide band running northeast to southwest across the cluster's face.

Now let's move over to the sting of the Scorpion, marked by the bright stars Lambda (λ) and Upsilon (υ) Scorpii. But since this is the Scorpion's most dangerous feature we won't get too close here. A nice, dark night shows us a fuzzy spot visible to the unaided eye 4° left and a little higher than the sting. This is the beautiful open cluster **M7.** It shows individual stars even in a finder-scope. With a 2.4-inch scope at least 30 may be counted, the brightest being arranged in the shape of a # symbol with every other extension chopped off. The star at the end of the southwest extension has an orange hue. A 4-inch scope shows up to 100 stars here, including many in a fainter halo that swells the group to a diameter of 1.3°. A low-power, wide-angle eyepiece is needed to take in the entire cluster.

Another pretty open cluster can be found 4° northwest of M7. **M6** is fainter and smaller than M7, but under dark skies it may also be glimpsed with the naked eye. About 40 of its stars can be counted through a 3-inch telescope at low power. This is often called the Butterfly Cluster, for a little imagination helps us picture the outline of a butterfly with wings spread wide.

The orange star at the eastern edge of this cluster is **BM Scorpii,** which is *usually* the brightest star in M6. It actually varies from 5th to 7th magnitude in a cycle lasting just over two years. When faint, it must share the title of brightest cluster member with three other 7th-magnitude stars.

I hope you enjoy your close encounter with the Scorpion. Even more wonders lie next door in the constellaion Sagittarius, and I will describe some of them starting on page 110.

In the Lair of the Scorpion

Object	Type	Mag.	Dist. (l-y)	RA	Dec.
ν Sco	Double star	4.0, 6.3	440	16ʰ 12.0ᵐ	−19° 28′
β Sco	Double star	2.6, 4.9	530	16ʰ 05.4ᵐ	−19° 48′
ζ Sco	Optical double	3.6, 4.7	150, 6,000	16ʰ 54.6ᵐ	−42° 22′
BM Sco	Variable star	5.5–7.0	2,000	17ʰ 41.0ᵐ	−32° 13′
M4	Globular cluster	5.9	7,000	16ʰ 23.6ᵐ	−26° 32′
M80	Globular cluster	7.2	32,000	16ʰ 17.0ᵐ	−22° 59′
NGC 6231	Open cluster	2.6	6,000	16ʰ 54.0ᵐ	−41° 48′
Trumpler 24	Open cluster	—	5,200	16ʰ 57.0ᵐ	−40° 40′
M7	Open cluster	3.3	800	17ʰ 53.9ᵐ	−34° 49′
M6	Open cluster	4.2	2,000	17ʰ 40.1ᵐ	−32° 13′

Around the Bend in Draco

DRACO, THE DRAGON, twists tortuously through the northern sky, wrapping himself around neighboring polar star patterns. You can pick up the tip of the Dragon's tail between the bowls of the Big and Little Dippers and follow along as he curls around the Little Dipper and then takes a sharp bend toward the south. A number of noteworthy sights hug the bend of Draco, making them fairly simple to ferret out.

We'll begin at the distinctive quadrilateral that marks Draco's head. Its faintest star is **Nu (ν) Draconis,** a double that is wide enough to split in binoculars. My 4.1-inch (105-millimeter) refractor at 17× shows matched suns aligned northwest (ν¹) to southeast (ν²). Both stars appear white to me, but there's just a hint of yellow in ν².

The star joining Draco's head to his sinuous body is Xi (ξ) Draconis, whose popular name Grumium ("jaw") sounds as if it came from a Dr. Seuss book. Drawing a line from Nu Draconis through Grumium and continuing for 1⅓ times that distance again will bring you to 5th-magnitude **39 Draconis.** At 17× I see an 8th-magnitude companion to the north-northeast, well split from the white primary. Boosting the magnification to 153× reveals a third component barely separated from the primary. Although both companions are the same brightness, the outlying one seems colorless, while the inner looks decidedly blue.

Moving 3.5° east-northeast of 39 Draconis takes us to slightly brighter **Omicron (o) Draconis.** Omicron is an unequal pair of orange stars easily seen at 17×. The 5th-magnitude primary harbors an 8th-magnitude sec-

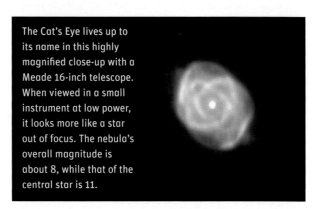

The Cat's Eye lives up to its name in this highly magnified close-up with a Meade 16-inch telescope. When viewed in a small instrument at low power, it looks more like a star out of focus. The nebula's overall magnitude is about 8, while that of the central star is 11.

ondary to the northwest. In his classic 1844 *Bedford Catalogue,* William H. Smyth described the pair as orange-yellow and lilac. Other observers have seen the companion as blue-green, but these hues must be a contrast illusion since the star has a spectral type of K3 (orange).

Up near the top of Draco's bend is Tau (τ) Draconis, an orange 4th-magnitude star. Sweeping 3.2° north from Tau, we come to a little clump of stars, one of which looks markedly reddish. This is the carbon star **UX Draconis,** one of the reddest stars in the sky. Continuing our discussion of star tints from page 92, the white star Vega (in Lyra) has a color index of 0.0, while reddish-orange Antares (in Scorpius) has a color index of 1.8. Yet UX Draconis is much ruddier than Antares, with a color index near 2.7. It is a semiregular variable that shows changes in both color and brightness. UX Draconis goes from around magnitude 5.9 to 7.1 and back over a cycle of about 175 days.

Next we'll seek out the charming asterism **Kemble 2.** This handful of 7th- to 9th-magnitude stars is very easy to locate, about 20′ across and centered just 1.1° east-southeast of 3.6-magnitude Chi (χ) Draconis. The name Kemble 2, which appears in the second edition of *Uranometria 2000.0* and the software atlas *MegaStar 5,* honors Lucian J. Kemble (1922–99), a well-known Canadian amateur and Franciscan friar who had written about the asterism in an article he never published. The description was brought to the attention of observers by his Norwegian friend, Arild Moland, who dubbed it the Mini-Cas asterism. In *The Deep Sky: An Introduction,* Philip S. Harrington calls this striking bunch the Little Queen.

These alternate names echo what is so striking about Kemble 2, an uncanny resemblance to the familiar **W** of the five bright stars of Cassiopeia, the Queen. There's even a sixth star corresponding to Eta (η) Cassiopeiae, albeit slightly out of place. The shape is nicely displayed in the *Millennium Star Atlas* but not labeled. Through my little refractor at 28×, the brightest stars appear golden, while the northernmost looks or-

A spiral galaxy seen nearly edge on, NGC 6503 looks more like a spindle than a spinning wheel. While it is bright enough to be spotted in small scopes, you won't see the fine structure visible in this image by Adam Block, taken on Kitt Peak in Arizona. North is up.

ange. Arresting though it may be, Kemble 2 is a chance alignment of stars lying at different distances and moving in different directions through space.

Scanning 4° westward we come to another wide double, **Psi (ψ) Draconis.** It is well split through my little scope at 17×. Both appear pale yellow, with the secondary north-northeast of the primary.

Psi Draconis makes a nice jumping-off point for the only extragalactic object in our eclectic list at right, the dwarf spiral galaxy **NGC 6503.** To locate it, put Psi at the western edge of a low-power field and sweep 2.1° south. When I recently viewed it low in a slightly hazy sky, my 4.1-inch scope at 68× showed a lens-shaped glow that brightened slightly toward the long axis (running west-northwest to east-southeast). The galaxy appeared about 3′ long and one-quarter as wide. An 8.6-magnitude star lies just east of the southern end. This high-surface-brightness galaxy was discovered in 1854 by a German university student, Arthur von Auwers, with his 2.6-inch refractor. Auwers later became a professional astronomer, and a lunar crater bears his name.

Last but not least, we'll visit **NGC 6543,** the Cat's Eye (Caldwell 6). This minute planetary nebula sits 3.6° south and a little east of NGC 6503. You can find it halfway between the 5th-magnitude stars Omega (ω) and 36 Draconis. At 17× it appears almost stellar, but its telltale robin's-egg-blue color gives it away. The nebula's central star can be distinguished even at this low power. A 10th-magnitude star just 2.7′ to the west-northwest keeps it company. At 153× the Cat's Eye looks slightly oval (elongated north-northeast to south-southwest) and perhaps a little darker in the very center. The outer edge seems to fuzz away into a thin and faint ring. The nebula takes magnification well if the air is steady.

In 1864 English amateur astronomer William Huggins examined NGC 6543 with a spectroscope. It was the first planetary nebula to be analyzed in this manner, and its spectrum proved that it is composed of luminous gases. (Prior to this, many astronomers had believed that with powerful enough instruments all nebulae could be resolved into stars.) Huggins could not identify the bright spectral emission lines that give the nebula its blue-green color, so they were attributed to a hypothetical new element named nebulium.

Later studies showed that these were the "forbidden lines" of doubly ionized oxygen. This type of emission is virtually nonexistent on Earth, but it can take place within the rarefied gases of some nebulae. Thus, as Robert Burnham Jr. says in *Burnham's Celestial Handbook,* "The forbidden lines, then, are not truly 'forbidden' at all; they are merely, so to speak, frowned upon severely. . . ."

The Dragon's Deep-Sky Delights

Object	Type	Mag.	Size/Sep.	Dist. (l-y)	RA	Dec.
ν Dra	Double star	4.9, 4.9	62″	99	17ʰ 32.2ᵐ	+55° 11′
39 Dra	Triple star	5.1, 8.0, 8.1	90″, 3.7″	190	18ʰ 23.9ᵐ	+58° 48′
o Dra	Double star	4.8, 8.3	37″	320	18ʰ 51.2ᵐ	+59° 23′
UX Dra	Carbon star	5.9–7.1	—	2,000	19ʰ 21.6ᵐ	+76° 34′
Kemble 2	Asterism	7 to 9	20′	—	18ʰ 35.0ᵐ	+72° 23′
ψ Dra	Double star	4.6, 5.6	30″	72	17ʰ 41.9ᵐ	+72° 09′
NGC 6503	Galaxy	10.2	7.1′ × 2.4′	17 million	17ʰ 49.5ᵐ	+70° 09′
NGC 6543	Planetary nebula	8.1	22″ × 19″	3,000	17ʰ 58.6ᵐ	+66° 38′

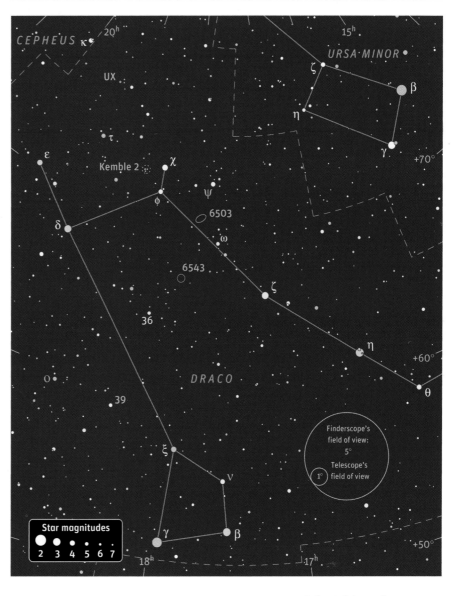

Neighbors

THE EASTERN LEG of Ophiuchus, the Serpent Bearer, is a straight line of three stars in the southern sky thrusting down between Scorpius and Sagittarius (see the all-sky chart on page 26 and the finder chart at right). Eta (η) Ophiuchi is its northernmost star, Theta (θ) Ophiuchi is the central one, and 45 Ophiuchi is the southernmost.

Ophiuchus's long limb cuts across rich star fields of the Milky Way and contains three of Messier's globular star clusters. But when following an observing list, we sometimes develop a sort of tunnel vision and fail to notice the sights nearby. This time we'll work our way through these clusters but pause to visit some of their lesser-known neighbors along the way.

M9 lies 3½° southeast of 2nd-magnitude Eta Ophiuchi. They'll fit together in the field of most good finderscopes, with 7.7-magnitude M9 showing as a small, dim, fuzzy spot not quite like a star. Through my 4.1-inch refractor at 153×, M9 appears fairly bright and about 3′ across with a large, slightly mottled, unresolved core and a fainter, granular halo. A few stellar pinpricks stand out in the halo, the most obvious one in the eastern edge.

Dropping the power to 17× reveals a conspicuous starless void 0.6° across: the dark nebula Barnard 64. It curves southwest from M9, then northwest. My little

Above: M19 and two other globular clusters including NGC 6284 at magnitude 8.8. The field is 2½° wide. North is up and east is left. *Right:* A close-up of M19, taken with a 5-inch f/8 reflector.

refractor with that 17× eyepiece gives a very wide field 3.6° across. This nicely encompasses not only M9 and B64 but also two dimmer globulars and another dark nebula as well. If your lowest-power eyepiece gives a field of less than 2°, you'll have to settle for viewing these objects one or two at a time.

NGC 6356 lies 1.3° northeast of M9. It's smaller but still obvious at 17×, displaying a large, relatively bright core, a dimmer halo, and a bright stellar nucleus. At high powers it looks slightly oval.

NGC 6342, 1.2° south-southeast of M9, is dimmer. My 17× view reveals a very small, faint smudge just a little brighter in the center. High powers fail to resolve any cluster stars, but a faint foreground star glimmers close to the northwestern edge.

A large dark nebula, Barnard 259, starts just west of this cluster, heads north, then east, then south in a fat, lumpy arc about 1° long. Although traceable with a small scope in a dark sky, B259 is less well defined than the dark nebula near M9.

Now drop 2.6° south to the 4th-magnitude yellow-white star Xi (ξ) Ophiuchi. Continue another 3.9° to 3rd-magnitude Theta (θ). A low-power view shows blue-white Theta contrasting nicely with a 6th-magnitude orange star 7′ to its northwest.

From here we can move 2.2° southwest to the pretty

The globular cluster M9 and its fainter neighbors; see the text for a description of the two dark nebulae B64 and B259.

binary star **36 Ophiuchi.** My 4.1-inch scope at 47× shows a nearly matched golden pair just 5″ apart. For a more comfortable split, try powers of around 100×. These are K0 dwarfs only a quarter as luminous as the Sun. They orbit each other with a period of several hundred years.

Now move 1.2° west to sweep up the globular cluster **NGC 6293.** At 153× it presents a dim halo about 2′ across surrounding a large, brighter core. **M19** lies 1.7° farther west and a little north. In a low-power, wide-angle view of 2° or more, it makes a nice globular pair with NGC 6293. But M19 appears much brighter and more than twice as wide. At magnitude 7.2, it can be recognized as a small, fuzzy patch in a good finder.

At 153×, M19 looks very pretty with gradual brightening toward the core and several stars sparkling in the halo. The cluster is decidedly oval, elongated north-northeast to south-southwest. It is more elongated than any other globular cluster known, perhaps due to the tidal force at its position a mere 5,000 light-years from our galactic core.

If I managed to lose you during our hop to M19, you can find it simply by starting at Antares and sweeping 7.4° due east. It's ³/₅ of the way from Antares to Theta Ophiuchi.

If you don't have an equatorial mount, here's another way to make that eastward sweep. Center Antares in a low-power eyepiece. Leave your telescope alone and spend 33 minutes watching for meteors, learning your way around the constellations, or doing whatever else you choose. Then look in your eyepiece again. M19 should be very near the center of the field!

Having found M19 one way or another, travel 3.8° south to locate an even brighter globular, 6.6-magnitude **M62.** At low powers you can see a wide double star (7th-magnitude orange and 8th-magnitude yellow) just 0.4° north of the cluster. Through my refractor at 153×, M62 shows an extensive, faint halo about 5′ in diameter with a large, much brighter core. The core is offset to the southeast within the halo and looks very mottled. A number of faint stars are scattered unevenly throughout the halo, especially north and west. M62 is the easiest of

The chart shows north up and east to the left. To find north in your eyepiece, nudge your telescope slightly in the direction of Polaris; new sky enters the view from the north edge. If you're using a right-angle star diagonal at the eyepiece, you'll probably see a mirror image that will not match sky maps. Either mentally flip the star patterns east-west or remove the diagonal and view straight-through.

these globulars to resolve into stars, but remember to use high powers while looking for them.

Many other globulars, dark nebulae, and double stars litter this area of the sky. The huge *Millennium Star Atlas,* which shows a million stars to 11th magnitude, details the richness to explore in much greater depth and at a larger scale than our 8.5-magnitude map here. If you keep your eyes peeled, you might bump into some of the sights just by sweeping around — making your own pleasant discoveries on a star-strewn summer night.

Six Globulars and a Double Star

Object	Type	Mag.	Dist. (l-y)	RA	Dec.
M9	Globular cluster	7.7	24,000	17ʰ 19.2ᵐ	–18° 31′
NGC 6356	Globular cluster	8.4	54,000	17ʰ 23.6ᵐ	–17° 49′
NGC 6342	Globular cluster	9.8	40,000	17ʰ 21.2ᵐ	–19° 35′
36 Oph	Double star	5.1, 5.1	19	17ʰ 15.3ᵐ	–26° 36′
NGC 6293	Globular cluster	8.2	28,000	17ʰ 10.2ᵐ	–26° 35′
M19	Globular cluster	7.2	28,000	17ʰ 02.6ᵐ	–26° 16′
M62	Globular cluster	6.6	18,000	17ʰ 01.2ᵐ	–30° 07′

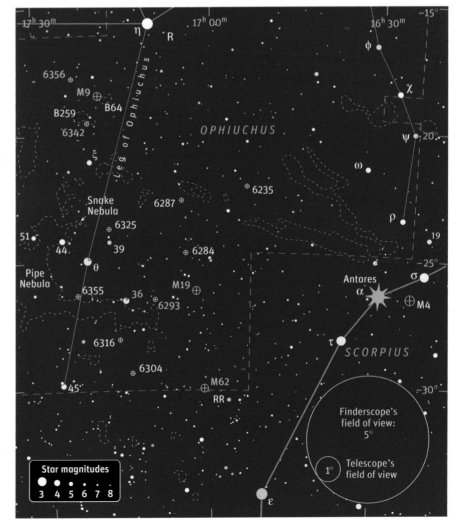

Overlooked in Ophiuchus

LARGE BUT DIM-STARRED Ophiuchus, the Serpent Bearer, is due south on July's all-sky chart on page 26. Mythologically, he is usually identified with the skilled healer Asclepius. In one version of the story, Asclepius witnessed one snake bringing another back to life by laying an herb on its head. Snatching the herb from the serpent, Asclepius secured this miraculous power for himself.

From his place in the sky, Ophiuchus still seems to practice the art of raising the dead. Our all-sky chart shows the healer standing over Scorpius, the scorpion that killed Orion. By about 1 a.m. near the end of July, Ophiuchus has trampled the loathsome arachnid into the southwestern horizon. Just before setting in the west the Serpent Bearer uses his healing herb on Orion, who flings his leg above the eastern horizon and rises into a sky freed of his mortal enemy.

Numeorus deep-sky sights within Ophiuchus are described elsewhere in this book (pages 86, 104, and 108), but many remain to be visited — among them, two Messier objects. We'll start from the naked-eye star Gamma (γ) Ophiuchi; from there, drop 1.4° south to the open cluster **Collinder 350.** With my 4.1-inch (105-millimeter) refractor at 17×, I see a large, loose group of about 35 stars within 50′. Most of the stars range from 9th to 11th magnitude and are arranged like four spidery arms.

Sweep just 50′ west of Gamma to 6th-magnitude **61 Ophiuchi.** Even at 17×, it is visible as a pretty pair of nearly matched stars aligned east to west. I see the eastern component as blue-white, the western as white.

From 61 Ophiuchi continue 1.5° west-southwest to another low-power double, **SHJ 251.** The golden, 6.4-

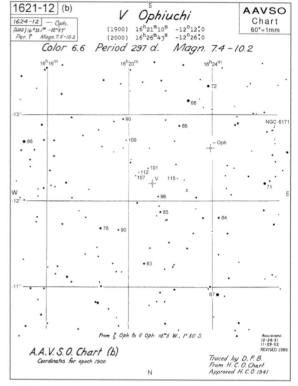

Check up on the brightness of V Ophiuchi by comparing it to surrounding "constant" stars. This chart gives their magnitudes to tenths, with the decimal point omitted. South is up. Hundreds of similar standard charts are available from the American Association of Variable Star Observers at http://charts.aavso.org/.

magnitude primary star has a 7.8-magnitude companion lying to its north-northwest. Because the secondary has a spectral class of F0, it ought to appear white to yellow-white, yet I have it logged as looking slightly orange.

Scanning 5.3° south from SHJ 251 brings us to the globular cluster **M14.** In 14 × 70 binoculars this cluster is small, round, and faint with a hint of brightening toward the center. My 4.1-inch scope at 153× shows a large core with a slightly dimmer halo about 5′ across. It appears granular, with some very faint pinpoints of light occasionally winking in and out of view.

If you find M14 easily, try for **NGC 6366,** a much tougher globular 3.1° to the southwest. It is centered just 16′ east of a 4.5-magnitude yellow-white star. NGC 6366 has been detected in scopes as small as 2.4 inches, but it has a very low surface brightness. No cluster members can be resolved in a small scope, but

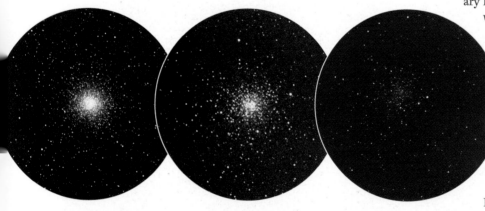

Not all globulars look alike! Back in 1966, Evered Kreimer captured densely packed M14 (left) and looser M107 (middle) from Prescott, Arizona, with his 12½-inch f/7 Newtonian reflector and Kodak Tri-X film cooled with dry ice. Martin C. Germano recorded sparse NGC 6366 (right) from Mount Pinos, California, with an 8-inch f/10 Celestron and gas-hypered Kodak Technical Pan 2415 film in 1987.

two faint stars (magnitudes 10.5 and 10.7) can be seen just beyond its western edge. Since it is fainter, you might expect NGC 6366 to be farther away than M14. But it is actually less than half the distance of its brighter companion.

Compare the finder chart below with the all-sky chart on page 26 if you need help locating the naked-eye star Zeta (ζ) Ophiuchi. From blue-white Zeta, just a 1.6° northward hop takes you to the open cluster **Dolidze 27.** My 4.1-inch scope at 68× shows eight stars forming a minimalist, Gemini-like stick figure. The western "head" star is brighter and yellow-white; the other "head" is yellow-orange. Four fainter stars, in a line running not quite east-west, form shared stick arms. Three stars south of these, also in a line, form the ends of stick legs with the center star shared. A dozen more stars (moderately bright to very faint) look as if they might be part of this very loose, 25'-wide group.

You can also use Zeta to find our second Messier object, **M107,** 2.7° south-southwest of the star. This globular cluster is just south of a colorful right triangle of stars, magnitudes 7.0, 7.4, and 7.6. From bright to dim they appear white, orange, and yellow. M107 and the star trio will fit together in a low-power field of view.

M107 is a nice sight through my little scope at 153×. It appears as a glow about 5' across with little brightening toward the center, but quite a few of its individual stars sparkle intermittently. Three stars of magnitude 11 or 12 cradle the cluster on its east-southeast, south-southwest, and west-northwest edges. They make a kite shape with a fourth star of similar brightness north of the globular.

M107 is about 21,000 light-years distant and 80 light-years across. It did not actually appear in Charles Messier's catalog. While compiling a bibliography of individual globular clusters in 1947, Canadian astronomer Helen Sawyer Hogg found a letter from Pierre Méchain in Johann Bode's *Astronomisches Jahrbuch* for 1786. In his letter Méchain mentioned four

"nebulae" that he'd discovered but had not passed on to Messier in time for inclusion in the catalog. Hogg claimed, logically, that these four objects should be appended to the Messier list with the designations M104 through M107.

The long-period variable star **V Ophiuchi** lies 1.5° west-northwest of M107. It typically ranges from magnitude 7.4 to 10.2 and back (with extremes of 7.3 and 11.6) over a period averaging 10 months. V Oph is a pulsating giant, and like the prototype of its class, Mira, it is distinctly reddish in color. Take a look at V Oph every now and then, with the help of a chart like that opposite, and you may be able to follow its rise and descent back into obscurity.

Some of Ophiuchus's Lesser Lights

Object	Type	Mag.	Size/Sep.	Dist. (l-y)	RA	Dec.
Cr 350	Open cluster	6.1	44'	—	17^h 48.1^m	+1° 18'
61 Oph	Double star	6.1, 6.5	21"	460	17^h 44.6^m	+2° 35'
SHJ 251	Double star	6.4, 7.8	111"	400	17^h 39.1^m	+2° 02'
M14	Globular cluster	7.6	11'	30,000	17^h 37.6^m	−3° 15'
NGC 6366	Globular cluster	9.2	13'	12,000	17^h 27.7^m	−5° 05'
Do 27	Open cluster	—	25'	—	16^h 36.5^m	−8° 56'
M107	Globular	7.9	13'	21,000	16^h 32.5^m	−13° 03'
V Oph	Variable star	7.3–11.6	—	900	16^h 26.7^m	−12° 26'

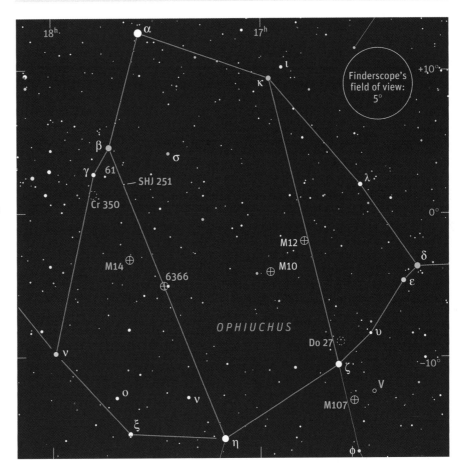

The Clusters of Taurus Poniatovii

MANY OLD STAR CHARTS show constellations that fell out of favor with astronomers and failed to make the list of 88 that have been officially recognized since 1930. One such is Taurus Poniatovii, Poniatowski's Bull, off the shoulder of Ophiuchus. It was created in 1777 by the Abbé Martin Poczobut, director of the Royal Observatory of Vilna, in honor of King Stanislaw Poniatowski of Poland. Although he imagined this miniature Taurus to be made up of just a small handful of stars, it encompassed dozens (mostly invisible to the unaided eye) when it appeared on Johann Bode's monumental *Uranographia* atlas in 1801. Taurus Poniatovii fell into disuse by the end of the 19th century

On our all-sky map on page 27, you can still locate the area once occupied by this constellation — between the constellations Serpens Cauda and Ophiuchus. Look about half-way up from the "Facing South" label for a tiny triangle of stars just above the "N" in Serpens Cauda.

The triangle is part of a **V** of five stars ranging from magnitude 3.9 to 5.7. Its likeness to the **V**-shaped Hyades group forming the face of the winter constellation Taurus inspired Poczobut to place this bull in the summer sky. Our close-up chart opposite shows that the **V** is formed by 66, 67, 68, 70, and 73 Ophiuchi. These stars are cataloged as part of the very large, very loose open cluster Collinder 359. But the existence of Cr 359 as a true cluster has been in doubt for some time, and data from the Hipparcos satellite indicate that this is merely a

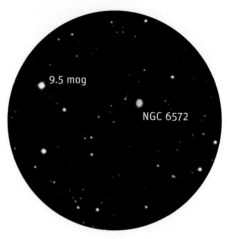

The little oval planetary nebula NGC 6572 appears almost starlike until you use high power on it. Compare its appearance to the 9.5-magnitude star 4′ east. Some observers describe the nebula as blue, others as green.

chance alignment of stars at different distances from us. The group is about 4° across, making it a target for binoculars rather than a telescope. Two of the stars in the V, however, are great doubles for a small scope.

67 Ophiuchi is a nice pair of blue-white stars. The 4th-magnitude primary has an 8.6-magnitude companion 55″ to the southeast. They can easily be separated even at very low powers. **70 Ophiuchi** is a little more challenging. Its components range from 1.5″ to 6.8″ apart with an orbital period of 88 years. Currently the stars are separated by 5.0″ and may require magnifications of 100× or more to split cleanly. These 4th- and 6th-magnitude gems are yellow-orange and orange, respectively, but color impressions are very subjective. Observers have described them as yellow and red, golden and rusty orange, or even pale topaz and violet. What colors do you see?

For our next target we'll start at bright Beta (β) Ophiuchi, also named Cebalrai. The big, loose open cluster **IC 4665** lies just 1.3° to its north-northeast. My 4.1-inch refractor at 47× reveals a coarse group of a dozen fairly bright stars and twice as many dim ones in an area 40′ across. Most of the brighter stars seem to be arranged in arcs. IC 4665 is only about 40 million years old, a young cluster dominated by hot stars in subtle shades of blue-white, white, and yellow-white. One exception lies off

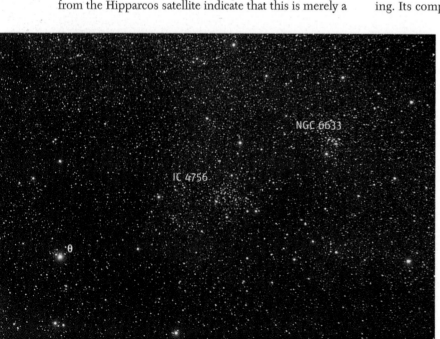

Compare this wide-field view with the star chart on the facing page; the star is Theta (θ) Serpentis.

the northeastern edge: an 8th-magnitude orange giant.

The wide double **Σ2216** lies in the eastern side of the cluster and is easily split at low powers. The 8th-magnitude primary appears white, but the yellow-white tint of the 10th-magnitude secondary is difficult to discern in a small scope.

Now let's push on to the more splashy cluster **NGC 6633.** With your lowest-power eyepiece, move 1° north of IC 4665 and then scan eastward for 10° (41 minutes of right ascension). You can also locate NGC 6633 by noting the nearly isosceles triangle it forms with the 4th-magnitude stars 70 and 72 Ophiuchi. With my 4.1-inch scope at 47×, I see about 30 stars of magnitudes 8 through 10 arranged in a wide bar 45′ long running northeast to southwest. A dozen more fairly bright stars lie outside the bar, and another 30 faint to very faint stars are scattered throughout this pretty cluster. A 5.7-magnitude blue-white star lies outside its south-southeastern edge. NGC 6633 is approximately 10 times older than IC 4665, and a few evolved, orange-hued stars may be seen among its members.

For our final cluster, move 3° east-southeast from NGC 6633 to find the very large group **IC 4756.** This object is a good binocular target, best viewed with a low-power, wide-angle instrument. My 4.1-inch scope with its 47× eyepiece encompasses a generous patch of sky 1.7° across, which nicely shows off this 1° cluster. In a more typical low-power view IC 4756 fills the field; you may not recognize it because you're looking right through it! Only a wider view will let you pick it out as a denser concentration against the rich Milky Way background. My 47× view shows a very rich, loose gathering of 100 moderately bright to faint stars. The cluster is slightly elongated east-west. Several brighter stars are sprinkled around the edges.

A pair of white stars whimsically known as Tweedledee and Tweedledum lie 1.6° due east of IC 4756. More officially known as **Σ2375,** the 6.4-magnitude primary has a 6.7-magnitude secondary 2.5″ to the east-southeast. Try powers of 150× or more to separate this nearly matched white pair. In 1953 the great double-star astronomer W. S. Finsen found both components to be extremely close pairs (much too close to see with a small scope). At the time, both pairs had nearly identical separations and orientations. Struck by this coincidence, he called them Tweedledee and Tweedledum after the two almost identical brothers in Lewis Carroll's classic *Through the Looking-Glass.*

Backtracking westward a way, you can hop among faint stars on our finder chart to locate the pretty little 8th-magnitude planetary nebula **NGC 6572.** The stars 72 and 71 Ophiuchi make a good starting place. Continue the line they make due south for 2° to a zigzag row of four 8th-magnitude stars ½° long. The nebula is 1° east of the zigzag's northern star. Its tiny greenish-blue disk is about 15″ across, slightly elongated north-south. At low powers it's practically starlike.

Sights in a Defunct Constellation

Object	Type	Size/Sep.	Mag.	Dist. (l-y)	RA	Dec.
IC 4665	Open cluster	0.7°	4.2	1,400	17ʰ 46ᵐ	+5° 40′
Σ2216	Double star	27″	8.0, 10.1	—	17ʰ 47.0ᵐ	+5° 42′
67 Oph	Double star	55″	4.0, 8.6	2,000	18ʰ 00.6ᵐ	+2° 56′
70 Oph	Double star	5.0″	4.2, 6.0	17	18ʰ 05.5ᵐ	+2° 30′
NGC 6572	Planetary nebula	15″	8	2,000	18ʰ 12.1ᵐ	+6° 51′
NGC 6633	Open cluster	0.8°	4.6	1,000	18ʰ 27.7ᵐ	+6° 34′
IC 4756	Open cluster	1°	5.4	1,300	18ʰ 39.0ᵐ	+5° 27′
Σ2375	Double star	2.5″	6.4, 6.7	700	18ʰ 45.5ᵐ	+5° 30′

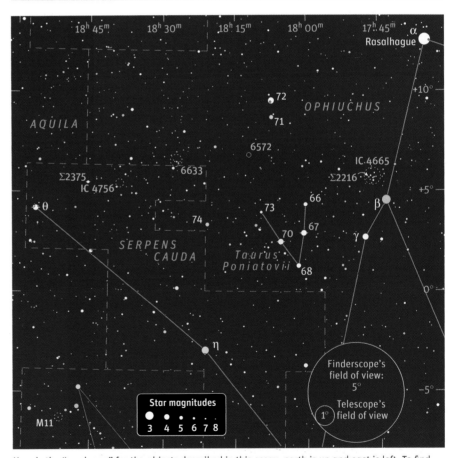

Here is the "road map" for the objects described in this essay; north is up and east is left. To find which way is north in your eyepiece, nudge your telescope slightly in the direction of Polaris; new sky enters the view from the north edge. Turn the map around to match. If you're using a right-angle star diagonal at the eyepiece you'll probably see a mirror image, which won't match sky maps. Either mentally flip the star patterns east-west or take out the diagonal and view straight through.

A Summer Steam Bath

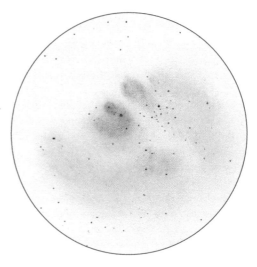

THE ANCIENT GREEKS thought of Sagittarius, the Archer, as a centaur — a creature with the body of a horse and the torso of a man. The Archer is often depicted pulling back on his bow with the arrow aimed at nearby Scorpius, the Scorpion. But this fanciful picture is difficult to imagine among the stars. Much more recognizable is the Sagittarius Teapot pattern. It's outlined on the all-sky chart on page 27 (where it's near the southern horizon) and in greater detail here.

Just above and left of the Teapot are four stars that we can make into a teaspoon. Below the Teapot there is even a lemon wedge, the brightest stars of Corona Australis. Completing the picture, the hazy band of the summer Milky Way stretches above Sagittarius like steam rising from our celestial Teapot's spout.

Let's immerse ourselves in the steam — for here is where some of the deep sky's greatest treasures reside. We'll concentrate on four of the brightest, arranged in a pattern like the corners of a slightly lopsided box (as marked on the chart at right).

Our first target is the Lagoon Nebula, **Messier 8,** located above the Teapot's spout. Under moderately dark skies M8 is easily visible to the unaided eye, looking like a small bright patch in this section of the Milky Way. Even a casual glance through a small telescope reveals a smattering of stars held in the embrace of a luminous

My drawing gives a good idea of how the Lagoon Nebula (M8) looks in my 4.1-inch refractor with a wide-field eyepiece. South is up and west is left to match the inverted view in a typical Northern Hemisphere telescope. The original pencil sketch *(above)* shows a "negative" view. The drawing was scanned and reversed in a computer to convey the eyepiece impression *(facing page)*.

cloud. Careful scrutiny with just a 60-millimeter (2.4-inch) scope will begin to show the dark lane that gives the Lagoon Nebula its name. The lane runs through the heart of the nebula and becomes more apparent in a 4-inch scope, along with a few dozen more stars.

No matter what telescope you have, two things will greatly improve the view: dark skies and experience. If you observe often, you'll gradually find yourself picking out more detail and detecting fainter nebulosity as your brain and eye learn to work together to make the most of the dim views in the eyepiece. Taking notes while observing speeds up this process. Writing a description of what you see compels you to look more carefully. Deep-sky observers also cultivate a technique known as *averted vision,* meaning that you look off to one side of an object instead of directly at it. This places the feeble glow on a more sensitive area of the eye's retina, making it easier to see.

Dark skies away from city lights make a dramatic difference. It is much easier to detect a basically gray nebula if you see it against a truly black sky. If you live under artificial skyglow, as most people do, a light-pollution-reduction (LPR) filter may help a little, making a nebula like the Lagoon stand out better. But a filter is a filter, and the embedded star cluster will appear a little dimmer.

I made the sketch of the Lagoon (above) with a 4.1-inch refractor under semirural skies, using an eyepiece that took in just under 1° of sky (two Moon widths). Notice that the nebula is brightest toward the south; this is the part you will find easiest to observe. A fainter swath of nebulosity curves around it to the east, north, and northwest but may take some time and patience to see.

After you have basked in the Lagoon a while, use the chart (right) to work your way leftward among the stars to the object labeled **M22.** This is one of the most

The four star clusters described in this section look like tiny clumps in this photo by Akira Fujii, much as they do to the unaided eye in the very darkest skies.

impressive globular star clusters in the sky. If it were farther north it would be better known, for it contains a half million stars seemingly packed into a dense ball. Even a 2.4-inch scope will show its hazy core surrounded by individual stars around the edges.

A careful look shows that this cluster is slightly oval. The view is very pretty through my 4.1-inch scope, for some stars can be resolved right across the center — reminding me of a diamond crushed to dust with surviving chips gleaming within the powder. The cluster's core may look dense to us, but in reality its stars are separated by a few *trillion* miles. Even so, on a planet near the center of M22 there would be no true night. The sky would be ablaze with stars brighter than any seen in Earth's night sky; some might even shine as brightly as our Moon!

In several ways M22 makes quite a contrast to the Lagoon Nebula. The Lagoon is a site of ongoing starbirth and is thought to be just 2 million years old. M22, on the other hand, ceased star formation long ago and is roughly 12 billion years old. M22 is one of the nearest of all globulars to our Sun. It lies at a distance of 10,000 light-years, but that's still twice as far away as the Lagoon.

Moving about 5° north of M22 (follow the map carefully!) we come to the pretty star cluster **M25.** A 2.4-inch scope shows a mixture of around 30 bright and faint stars in a loose, scattered group ½° across, so this object has all the earmarks of an open (as opposed to globular) cluster. Seven stars of 9th and 10th magnitude near its center trace the shape of a small, open-mouthed smile or capital letter **D**. The variable star **U Sagittarii** lies just east of this **D**. It is a Cepheid variable, a yellow giant that swells and shrinks with a precisely regular period — in this case once a week. U Sagittarii takes a little less than five days to fade from peak brightness (magnitude

6.3) to its dimmest (7.2); the rise back to 6.3 takes only two days. When at minimum light U Sgr is the third-brightest star in M25; at maximum it is the brightest.

Now let's shift 8° westward to **M23,** another open cluster. Its stars are much more uniform in brightness than those of M25. A 2.4-inch scope will reveal around 40 stars; a 3.5-inch can increase the count to more than 50. The whole group appears oblong in the northeast-southwest direction, and many of its stars seem arranged in spidery arms. M23's brightest member is an 8th-magnitude yellow star near the cluster's north-eastern edge. A brighter, white star lies just outside the cluster to the northwest.

These sights by no means exhaust the riches in the mist above our Teapot. But they do make a good place for novice telescope owners to get their feet wet — or steamy, as the case may be.

Misty Splendor in Sagittarius

Object	Type	Mag.	Dist. (l-y)	RA	Dec.
U Sgr	Variable star	6.3–7.2	2,000	18h 31.9m	–19° 08'
M8	Nebula/open cluster	5.8	4,000	18h 03.8m	–24° 23'
M22	Globular cluster	5.1	10,000	18h 36.4m	–23° 54'
M25	Open cluster	4.6	2,000	18h 31.6m	–19° 15'
M23	Open cluster	5.5	2,000	17h 56.8m	–19° 01'

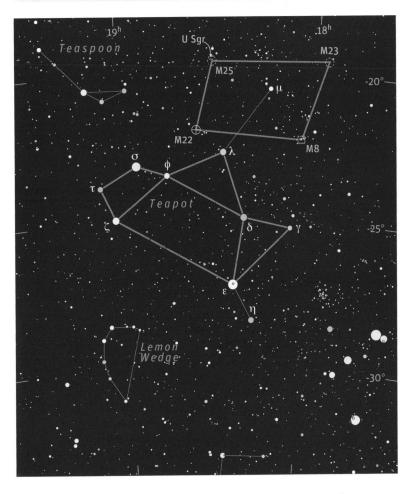

Low in the southern sky, the Teapot of Sagittarius looks about as big as your hand at arm's length. This chart includes stars to magnitude 8.0, about what can be seen in good binoculars.

Floating on Cloud 24

FOR MANY PEOPLE Cloud Nine may be the site of their heavenly moments, but those of us with our heads in the stars can reach a bit higher — to Cloud 24 and the deep-sky mists floating above it.

As a Messier object, **M24** is in a class by itself. It is not a true star cluster but rather a rich star cloud in one of our galaxy's spiral arms. The hazy band of the Milky Way rises from Sagittarius like steam from the constellation's teapot-shaped asterism. If you are fortunate enough to have dark skies, you can see this misty river of light and the dark rift running through it. For the most part, the stars and nebulae that lie along this section of the Milky Way, as well as the dark clouds that make up its rift, lie within the Sagittarius Arm of our galaxy. Within the dark rift, however, there is a gap in the Arm that allows us to see deep into the galaxy. Through that hole shines M24, the Small Sagittarius Star Cloud.

Under dark skies you can see it with the unaided eye just north of 3.8-magnitude Mu (μ) Sagittarii. It measures about 1° × 2° and is nearly rectangular, with the long dimension running northeast to southwest. M24 is best appreciated in binoculars or a small, rich-field telescope at low power giving a true field of 2° or more. The stars range from a relatively bright 6th magnitude down to the faintest pinpricks of light your telescope

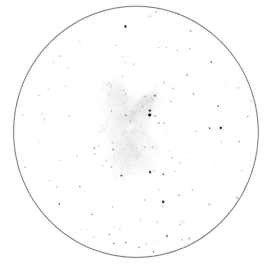

The beauty of star cluster M16 and the Eagle Nebula (IC 4703) is apparent through both large and small telescopes. *Above:* My sketch shows how they look through a small scope at 87×. *Facing page:* Astrophotographer Jose Torres captured their dramatic appearance through a 10-inch telescope using a CCD camera.

can pick out. Try scanning the area with your lowest-power eyepiece to appreciate the impressive wealth of stars that adorns this beautiful corner of the sky.

M24 has a number of dark nebulae within its borders. The most obvious is **Barnard 92,** which lies near the edge about halfway along the northwestern side. It is roughly oval, about 1/4° long by half as wide, and elongated north to south with the eastern side having the most sharply defined edge. A 4-inch scope shows a lone star shining within. If you don't find B92 too difficult, try looking for neighboring **Barnard 93,** located 1/3° to the east-northeast. B93 is also aligned north-south and is about as long as B92 but only half the width. B93 has less sharply defined borders than B92, making it a more difficult target. These nebulae give M24 the comical appearance of two dark eyes in a fuzzy face.

The small, sparse open cluster **M18** lies about 2/3° above the northern edge of the Small Sagittarius Star Cloud. Often bypassed in favor of its more conspicuous neighbors, M18 is visible as a small, hazy patch through binoculars. A 3.6-inch scope shows a dozen glittering stars ranging from magnitude 8.7 to 10.6 in a loose cluster about 8′ across. The three brightest of these form an arc in the northern part of the cluster. Just 2.5′ south of the arc's middle star is a wide double that can be split at 30×. Low powers lend M18 a more concentrated, cluster-like appearance, but high magnifications may bring out some of the

The Sagittarius sector of the Milky Way holds many starclouds and nebulae. Near the center of this image of M24 you'll find two dark nebulae, Barnard 92 (right) and 93 (left).

near the heart of the nebula. This is the only dark feature that can be glimpsed in a small telescope, and it's extremely difficult to detect. The Eagle Nebula is one of the few well-known observing targets that respond well to an H-β (hydrogen-beta) light-pollution filter, but that filter makes the view too dim on scopes smaller than 6 inches.

So after having our heads in a (star) cloud, it is only appropriate that we next go after some of the lesser-known sights elsewhere in Sagittarius.

faint stars in both the cluster and the background Milky Way.

M17 is a hazy patch is variously known as the Omega Nebula, the Swan Nebula, the Checkmark Nebula, the Horseshoe Nebula, and the Lobster Nebula. Such are the vagaries of human imagination. The brightest part of the nebula has always reminded me of a swan or the number 2 with a long bar. Since the name 2 Nebula lacks any poetic charm, I like to call M17 the Swan Nebula. Look for it 58′ north-northeast of M18. It is visible through a small finder as a little hazy spot. Through my 4.1-inch refractor at 87× the body of the swan appears mottled and runs west-northwest to east-southeast. The swan's neck stretches southward from the western end of the body, and her head curves back toward the west with a star embedded in the end of her beak. A small circle of nebulosity hovers over her head. Some extremely faint nebulosity can also be seen above the swan's back and behind her tail. Under mild light pollution, these dim outer portions may be seen more easily with a narrowband filter.

The open cluster **M16** is embedded in a nebula (IC 4703). It's often called the Eagle or Star Queen Nebula. As a novice observer, I sometimes found M16 easier to spot in my finder than through my telescope. M16 is located 2.4° north-northwest of M17, and the pair fits in the same field through most finders. Through my 4.1-inch scope at 87×, about 30 stars can be seen scattered across the nebula, but only those reaching from the center to the northwest are generally considered part of the cluster. The nebulosity is faint and distributed in three arms — one reaching toward the northeast, one toward the northwest, and a very wide one toward the southwest. Photographs show that the Eagle contains many dark nebulae. The most prominent is a backward **L** of darkness, the Star Queen's throne,

On this chart north is up and east is left. The circles show fields of view for a typical finderscope or a telescope with a low-power eyepiece. To find north through your eyepiece, nudge your telescope toward Polaris; new sky enters the view from the north edge. (If you're using a right-angle star diagonal it probably gives a mirror image. Take it out to see an image matching the map.)

Exploring a Star Cloud

Object	Type	Mag.	Dist. (l-y)	RA	Dec.
M24	Star cloud	4.5	9,400	18ʰ 16.9ᵐ	−18° 29′
B92	Dark nebula	—	—	18ʰ 15.5ᵐ	−18° 11′
B93	Dark nebula	—	—	18ʰ 16.9ᵐ	−18° 04′
M18	Open cluster	6.9	3,900	18ʰ 19.9ᵐ	−17° 08′
M17	Nebula	6.0	4,900	18ʰ 20.8ᵐ	−16° 11′
M16	Open cluster/nebula	6.0	8,200	18ʰ 18.8ᵐ	−13° 47′

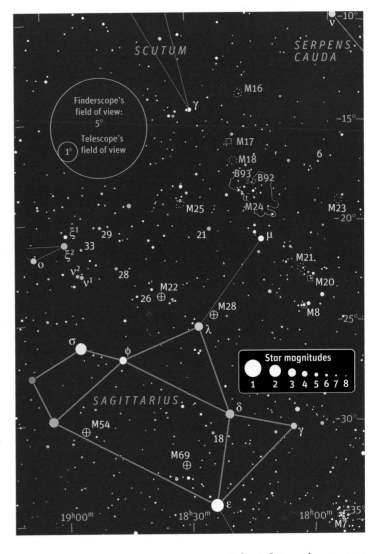

After Tea

THE EYE-CATCHING Teapot outlined by the brightest stars of Sagittarius sits low in the south on this month's all-sky chart on page 27. The Milky Way rises above it, like steam from the Teapot's spout, and captivates observers with its deep-sky largesse. Rapt skygazers seldom turn their attention to the barren-looking region that follows the Teapot in its westward march across the sky. Yet this area holds two Messier objects and other intriguing sights waiting to be added to your list of small-scope captures.

Our targets can be a challenge to find in the empty, star-poor regions of eastern Sagittarius — so we'll begin with the globular cluster **M55,** which has the redeeming quality of being visible through a finderscope as a little fluff ball. Here are a few tactics to get you in the correct area. Visualize a line from Sigma (σ) through Tau (τ) Sagittarii (the end stars in the Teapot's handle) and continue for 2³/₄ times that distance again. Also note that M55 makes a long isosceles triangle with the handle stars Tau and Zeta (ζ).

You can start a bit closer if your sky is dark enough to spot 52 and 62 Sagittarii, two stars that make a nearly equilateral triangle with M55. Put 52 Sagittarii at the western edge of a low-power field and then sweep 6°

Because it is relatively close to our own Milky Way system, NGC 6822 looks more like a telescopic version of a Magellanic Cloud than a typical galaxy. The bright star ³/₄° to its south-southwest (lower right) is magnitude 5.5. Martin C. Germano hauled his 8-inch f/5 reflector up Mount Pinos in California for this photograph on hypered Kodak 2415 emulsion.

southward. With a little luck, any of these methods should get you close enough to recognize M55 through a finder or low-power eyepiece.

With a 2.4-inch (60-millimeter) telescope at around 50×, M55 appears large and mottled with a broad, slightly brighter core. A 4- to 6-inch scope at 75× to 100× will bring out many faint stars loosely scattered across the cluster. Including the sparse outer halo, the group seems about 10' across. Dark lanes fracture it, and the borders of the cluster are ragged with nibbles taken out of the edge, including a rather large one in the southeast side.

M55 is one of the closest globular clusters, approximately 17,000 light-years away and 100 light-years across. It shines with nearly 90,000 times the luminosity of our Sun.

Our second Messier object is the globular cluster **M75,** which hugs the eastern border of Sagittarius. It's easy to scan past without noticing, so you must precisely pin down its position. Start at 4.4-magnitude, reddish orange 62 Sagittarii, and place it outside the southern edge of a finder field to reveal a pair of

A floating ball of stars, M55 would be much better known if it could be seen high in the sky from north temperate latitudes. North is up in this 1° square view. Lee C. Coombs of Atascadero, California, used his 10-inch f/5 reflector for this 25-minute exposure on Kodak Ektachrome P1600 film.

6th-magnitude stars to the north. With a low-power eyepiece in your telescope, position the eastern member of the pair near the western edge of your field of view. This should put M75 near the northern perimeter. If you use a unit-power finder (an illuminated 1× targeting device), note that M75 lies about halfway between Psi (ψ) Capricorni and Rho (ρ) Sagittarii. You'll see these stars labeled on the chart below.

Through a small telescope at low power, M75 resembles a fuzzy star. My 4.1-inch refractor at 87× shows a little ball of haze about 2′ in diameter with no resolved stars. The cluster brightens sharply toward the center, where a tiny, intense nucleus resides.

At 68,000 light-years, M75 is one of the most remote Messier globulars, making Charles Messier's description quite curious. He said that M75 seemed to be composed only of very faint stars, containing some nebulosity — and yet Messier saw no stars in M55 or other bright globulars much easier to resolve. M75 appears small and faint only because of its great distance. It is about 130 light-years across and over 220,000 times more luminous than our Sun.

Now let's move to Rho Sagittarii at the northern end of an asterism known as the Teaspoon. **NGC 6774** lies just 2° northwest of this star. It is a large and sparse open cluster about 30′ across, best viewed through binoculars or a small scope at low power. It hides in a rich star field, and a wide view is needed to help it stand out from its background. My little refractor at 28× shows about 40 stars of magnitudes 8 through 12 arranged in little bunches and chains.

NGC 6774 shares the finder field with 4.5-magnitude Upsilon (υ) Sagittarii. Placing Upsilon a little way outside the western edge of the field will bring a curve of three 5th-magnitude stars into the eastern side. Moving from the middle star to the northern one and continuing for that distance again will take you to the location of **NGC 6822,** Barnard's Galaxy. It was discovered in 1884 by American astronomer Edward Emerson Barnard while he was still an amateur in Tennessee, but it was not recognized as an extragalactic object until the 1920s. Barnard's Galaxy (also called Caldwell 57) is a member of the Local Group, a small cluster of about 40 known galaxies including our own.

Barnard described his discovery as "an excessively faint nebula . . . very diffuse and even in its light. With 6 inch Equatoreal it is very difficult to see, with 5 inch and a power of 30± (field about 1¼°) it is quite distinct. This should be borne in mind in looking for it."

Barnard's advice is worth heeding. It's often easier to catch sight of NGC 6822 in a small scope that allows a wide field

than in a large scope with its more restricted view. While Barnard's Galaxy has been seen in 7 × 35 binoculars, I recommend at least a 2.4-inch telescope under fairly dark skies. With my 4.1-inch refractor at 17× from my semirural home the galaxy is elusive, appearing very faint and oblong. The long dimension measures about 11′ and runs north-south.

Barnard's report in the prestigious German journal *Astronomische Nachrichten* was entitled "New Nebula Near General Catalogue No. 4510." Today we know GC 4510 as **NGC 6818,** a small, bright planetary nebula nicknamed the Little Gem. It lies 41′ north-northwest of Barnard's Galaxy, and the two fit within the same low-power field. If you were unable to find Barnard's Galaxy, return to the curve of 6th-magnitude stars below it and sweep 1.3° north from the northernmost star.

Through my little refractor at 28×, the Little Gem looks like a bloated, bluish star. At 87× it can be recognized as a tiny, round nebula. At 153× the planetary hints at being oval with perhaps some slight patchiness. In a sky largely populated with ancient wonders, this aquamarine jewel is comparatively young — a mere 3,500 years old.

Sagittarius Sidelights

Object	Type	Mag.	Size/Sep.	Dist. (l-y)	RA	Dec.
M55	Globular cluster	6.3	19′	17,000	19^h 40.0^m	−30° 58′
M75	Globular cluster	8.5	6.8′	68,000	20^h 06.1^m	−21° 55′
NGC 6774	Open cluster	—	30′	820	19^h 16.6^m	−16° 16′
NGC 6822	Galaxy	8.8	16′ × 14′	1.6 million	19^h 44.9^m	−14° 48′
NGC 6818	Planetary nebula	9.3	22″ × 15″	5,500	19^h 44.0^m	−14° 09′

Angular sizes are from various catalogs, but NGC 6774 has such indefinite borders that size claims vary widely.

Night of the Trifid

DAYTIME TRIFFIDS may be the giant, man-eating plants of science-fiction fame, but the Trifid of the night is even more remarkable. The Trifid Nebula is an amazing complex, for it contains nebulae of three different types, an open star cluster, and a multiple star. But the best way to locate the Trifid Nebula is by first visiting its spectacular neighbor, the Lagoon, which I first described on page 110.

Look for a concentrated patch of mist above the spout of the Teapot asterism formed by the bright stars of Sagittarius. From my northerly location in upstate New York, the **Lagoon Nebula (M8)** sails low across the southern sky. Yet I can pick it out with my unaided eye even in suburban skies. Turning a small telescope toward M8, the first thing you'll notice is the embedded star cluster **NGC 6530.** With my 4.1-inch refractor at 87×,

Under the very best conditions, when the Trifid (upper center) and Lagoon (lower left) are high in a very dark sky, observers using small telescopes at a variety of magnifications can trace nearly all the nebulosity visible in this image. The main difference is that the human eye is almost completely colorblind in dim light — nebulae, like cats, look mostly gray at night.

I count 25 stars in a wedge-shaped group that points toward the northeast. The stars are 7th through 11th magnitude, and the longest dimension is about 10′ (one-third of the Moon's breadth in the sky). Two smaller gatherings, of several stars each, lie north and west of the main group. The latter contains the 6th-magnitude star 9 Sagittarii.

The nebula itself is designated **NGC 6523.** At first blush, you may notice nebulosity only near the concentrations of stars. The most obvious patch envelops 9 Sagittarii and includes a particularly bright knot 3′ to the star's west-southwest. When viewed at high magnifications, the knot's pinched shape inspired its popular name: the Hourglass. The 9.5-magnitude star alongside the Hourglass is its source of illumination. The second-brightest area of the Lagoon, involving the starry wedge, is best seen where it's smeared to the cluster's southwest. A slash of blacker sky runs between these glowing clouds; it's part of the dark nebula that gives the Lagoon its name.

A large but more diaphanous region of gossamer light spreads east and south from the cluster. Dark skies will make such faint sections come to life, but if your skies are light polluted you can improve the view considerably with the special filters available. The nebula responds very well to both narrowband and O III filters.

Look next for a dim haze engulfing the northern smattering of stars and extending westward. In my 4.1-inch scope at 87×, I can follow it for about 26′ to the 5th-magnitude star 7 Sagittarii. The nebula fades just before it reaches the double star **Argelander 31.** The pair's 7th-magnitude primary has a 9th-magnitude secondary 34″ north-northeast. This wide and gently curving swath of light cradles the brighter nebulosity around 9 Sagittarii, from which it is separated by another leg of the Lagoon's dark channel.

From the Lagoon, finding the **Trifid Nebula (M20)** is simple. Center the star 7 Sagittarii in a low-power field, scan slowly north for 1.3°, and look for a 7th-magnitude star surrounded by a tenuous haze. This is the glow of the nebula itself, which bears the designation **NGC 6514.** But the name Trifid comes from the fact that a dark nebula, **Barnard 85,** divides it into three lobes nearly centered on the 7th-magnitude star. B85 has been seen in scopes as small as 2.4 inches; even without a filter, it is easily visible under semirural skies in my husband's 3.6-inch refractor at 64×. When light pollution interferes, O III and narrowband filters can improve the view. In photos, B85 divides the nebula into four sections. Noted observer Stephen James O'Meara likens it to a four-leaf clover, since you'd be lucky to glimpse the fourth lobe through a small scope.

The central star of the Trifid is a multiple system with seven known components. My 4.1-inch scope at 153× nicely displays the two brightest stars. Designated **H N 6,** this pair has a 7.6-magnitude primary with an 8.7-magnitude secondary 11″ south-southwest. On a steady night you might spot a third member in a nearly straight line with the other two. Look for a faint 10.4-magnitude star 6″ north-northeast of the 7.6-magnitude star. This pair is known as **H N 40.**

It is the small group of hot stars at the heart of the Trifid that allows us to see the nebulosity surrounding them. The energy from these stars ionizes the hydrogen gas around them. When escaped electrons recombine with the hydrogen atoms, they emit the characteristic red light that is the hallmark of emission nebulae and is so stunning in photographs.

Photos also show a bluish nebula surrounding the 7th-magnitude star 8′ north-northwest of the triple. Here, smoke-size dust particles reflect blue light from the embedded star. The reflection nebula is visible in a small scope, though it seems fainter than its glowing cousin to the south. In dark, haze-free skies it looks nearly as large as the emission nebula, but in reality it's much larger. Human eyes cannot see the beautiful colors so striking in photographs, but if I look closely I get the impression that the parts have subtly different hues. I also find that filters do little to bring out the reflection nebula — they may even make it *less* apparent.

In many deep-sky references, M20 is listed as a "nebula + cluster." The many stars scattered in and around the nebula presumably belong to **Collinder 360.** The cluster is not conspicuous against the rich background of the Milky Way, but a small scope will show the area spangled with one or two dozen stars.

The open cluster **M21** lies 40′ northeast of the Trifid. The two bright stars of the Trifid and three additional 7th- and 8th-magnitude stars form a gentle arc that will lead you right to it. About 20 stars are gathered in a small knot with ill-defined borders; high magnifications will show more. The brightest star is the

double **S698.** It has a 7.2-magnitude primary with an 8.5-magnitude secondary 30″ northeast.

All the objects I've mentioned fit in a 2½° circle of sky, meaning that a small scope of short focal length could embrace them all in one field of view. Most lie within the amazing complex of M20. When gazing at the Trifid, remember that you're seeing an area where new stars are forming. Will one of them eventually warm a planet with huge, carnivorous plants — or with something still more wondrous and strange?

Trifid Nebula (M20) and Environs

Object	Type	Mag.	Size/Sep.	RA	Dec.
NGC 6514 (main part)	Emission nebula	—	16′	18ʰ 02.4ᵐ	−23° 02′
B85	Dark nebula	—	16′	18ʰ 02.4ᵐ	−23° 02′
NGC 6514 (north part)	Reflection nebula	—	20′	18ʰ 02.5ᵐ	−22° 54′
H N 6 and H N 40	Multiple star	7.6, 8.7, 10.4	11″, 6″	18ʰ 02.4ᵐ	−23° 02′
Collinder 360	Open cluster	6.3	13′	18ʰ 02.5ᵐ	−23° 00′
M21	Open cluster	5.9	12′	18ʰ 04.6ᵐ	−22° 30′
S698	Double star	7.2, 8.5	30″	18ʰ 04.2ᵐ	−22° 30′

Parts of the Lagoon Nebula (M8)

Object	Type	Mag.	Size/Sep.	RA	Dec.
NGC 6530	Open cluster	4.6	15′	18ʰ 04.8ᵐ	−24° 20′
NGC 6523	Emission nebula	5.8	90′ × 40′	18ʰ 03.8ᵐ	−24° 23′
Arg 31	Double star	6.9, 8.6	34″	18ʰ 02.6ᵐ	−24° 15′

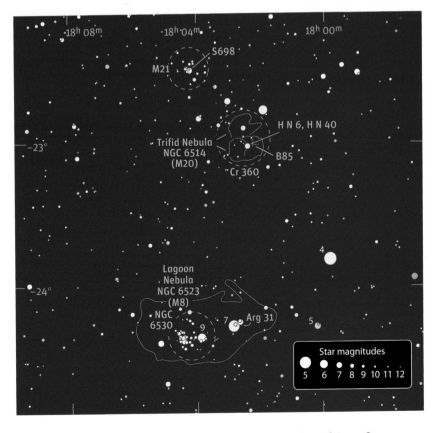

Tea Cozy

SAGITTARIUS, THE ARCHER, glides across the southern horizon on late summer evenings and tempts us with its deep-sky riches. Many lie near the bright stars that form the constellation's distinctive teapot-shaped asterism, so let's cozy up to the Teapot and see what we can find.

We'll start at the top of the Teapot, marked by golden-hued Lambda (λ) Sagittarii, also known as Kaus Borealis. As I noted on page 110, Sagittarius was originally depicted as a mythical centaur drawing his bow, the arrow seemingly aimed at nearby Scorpius. Kaus Borealis means the "northern (part of the) bow."

The magnificent globular cluster **M22** is centered just 2.4° northeast of Kaus Borealis, the two being easily visible in the same finder field. M22 is a gorgeous sight in my 4.1-inch (105-millimeter) refractor at 87×. It

shows a very large halo nearly 20′ across with an abrupt increase in star density halfway in. Stars can be resolved right across the bright 4′ core, which sparkles like a mound of glittering sugar. The cluster is adorned with many star chains, clumps, and dark patches, and it appears slightly elongated northeast to southwest. A 9th-magnitude star is visible at the outer edge of the halo to the north-northeast.

Globular clusters are relatively old and contain many yellow, orange, and reddish stars. Although I find the colors of globular clusters difficult to discern, some observers with more color-sensitive eyes can detect their warm hues. The spectrum of M22 resembles that of an *F*5 star, so it ought to shine with a yellow-white light. Can you distinguish its color?

Another globular cluster, **M28**, lies 1° northwest of Lambda Sagittarii. While M22 owes its glory to the fact that it is one of the nearest globulars, M28 is almost twice as far away at 18,300 light-years and appears correspondingly dimmer. It also looks smaller than the catalog size given in our table at right. At 87× a faint 5′ halo shows only a few stars surrounding a 2′ core. Boosting the power to 127× brings out a splash of stars scattered south and southeast of the core. Stretching north from the core, some stars at the edge of perception form two curved pincers. The overall spectral type of M28 is *F*8, giving it a pale yellow glow.

Now we'll move to yellow-orange Delta (δ) Sagittarii, which defines the spot where the lid and the spout of the Teapot meet. Delta is also known as Kaus Media, the "middle of the bow." **NGC 6624** is ³/₄° southeast of this star. It is the brightest non-Messier globular close to a Teapot star. NGC 6624 can't be resolved in a small telescope, but its tiny bright core makes it easy to spot. Observationally, the cluster is interesting mostly for its color. With a spectral type of *G*4/5, this is the yellowest of the globulars we'll visit.

Golden Gamma (γ) Sagittarii glitters at the business end of the Teapot's spout. As the tip of the Archer's arrow, it bears the common name Alnasl, "the point." Two attractive double stars can be found nearby.

Burnham 245 (often written β245 in books and BU 245 in computer databases) is 1° to the east-southeast of Alnasl. It's the brightest star in that area and consists

Each of these globular clusters is home to at least 30,000 stars, and in some cases perhaps 10 times that number. But M22, the nearest, is the only one in which many individual member stars can be seen visually in a small telescope. The photographs of the Messier objects were taken with telescopes of 8- to 12¹/₂-inch aperture; NGC 6624 is from the Digitized Sky Survey, Southern Hemisphere. All views are oriented with north up.

of a 5.8-magnitude yellow-orange primary with an 8.0-magnitude yellow companion 3.9″ north. They are comfortably split at 127×.

The second double is 1.5° west of Alnasl and the brightest star in the vicinity. At 87×, **Piazzi 6** is a pretty couple composed of a reddish orange 5.4-magnitude star with a yellow-orange attendant, one-fourth as bright, lying 5.5″ east-southeast.

At the southwest corner of the Teapot we find blue-white Epsilon (ε) Sagittarii, or Kaus Australis, "southern (part of the) bow." A gentle curve of three equally spaced 6th- and 7th-magnitude stars stretches between Kaus Australis and the 5.4-magnitude double star **Howe 43.** At 153× in my little telescope, the white primary has a faint companion close to the south.

The globular cluster **M69** sits ²/₃° north of Howe 43. At 127×, the cluster appears small (2′ across) and patchy, growing brighter toward the center. A few faint pinpricks of light glisten in the outer regions. An 8th-magnitude star lies outside the northwest edge. The cluster's integrated spectral type of G2/3 is similar to our Sun's, so M69 should have a slightly yellowish tint.

From M69, scan 2.5° due east to find **M70.** You can also use the Earth's rotation to help you find this globular. If you center M69 in the low-power field of an undriven telescope, M70 should be visible near the center about 12 minutes later. At 127×, M70 looks just a bit fainter and smaller than M69. A little curve of 10th-magnitude stars emanates from the northeast edge of the cluster. M70 has very nearly the same spectral type as M22.

The three globulars we've just seen lie 25,800 to 29,700 light-years away, which largely accounts for their lack of resolution in small telescopes. Our final target is the most distant of the bunch. **M54** lies an incredible 87,400 light-years away. Studies indicate that M54 and the fainter globulars Arp 2, Terzan 7, and Terzan 8 are part of the Sagittarius Dwarf Elliptical Galaxy, which is being cannibalized by our own Milky Way. In fact, M54 (by far the largest and brightest of the quartet) may actually be the nucleus of the hapless dwarf.

To find M54, start at the southeast corner of the Teapot, marked by Zeta (ζ) Sagittarii, and scan 1.7° west-southwest. This immensely distant cluster reveals only its inner region through a small telescope, and no member stars. Surprisingly, however, M54 looks a little brighter than M69 or M70 — a testament to the fact that it is among the most luminous globular clusters. At 127×, M54 displays a 1′

halo brightening smoothly toward a nearly stellar nucleus. Two faint foreground stars can be seen at the edge of the cluster, one to the east and the other to the southeast. With a spectral type of F7/8, M54 should have the same hue as M28.

The constellation Sagittarius is rich in globular clusters, including 14 brighter than magnitude 9.0. Using this essay and the star map below, you can get cozy with the bright stars of the Teapot and look in on nearly half of them. But there's more. If you haven't already done so, turn to pages 110, 112, and 114 to discover more wonders within this celestial archer.

Doubles and Globulars Near the Teapot

Object	Type	Mag.	Size/Sep.	Dist. (l-y)	RA	Dec.
M22	Globular cluster	5.1	24′	10,400	18ʰ 36.4ᵐ	−23° 54′
M28	Globular cluster	6.8	11′	18,300	18ʰ 24.5ᵐ	−24° 52′
NGC 6624	Globular cluster	7.9	6′	25,800	18ʰ 23.7ᵐ	−30° 22′
β245	Double star	5.8, 8.0	3.9″	360	18ʰ 10.1ᵐ	−30° 44′
Pz 6	Double star	5.4, 7.0	5.5″	900	17ʰ 59.1ᵐ	−30° 15′
Howe 43	Double star	5.4, 9.8	3.5″	350	18ʰ 31.1ᵐ	−32° 59′
M69	Globular cluster	7.6	7′	29,700	18ʰ 31.4ᵐ	−32° 21′
M70	Globular cluster	7.9	8′	29,400	18ʰ 43.2ᵐ	−32° 18′
M54	Globular cluster	7.6	9′	87,400	18ʰ 55.1ᵐ	−30° 29′

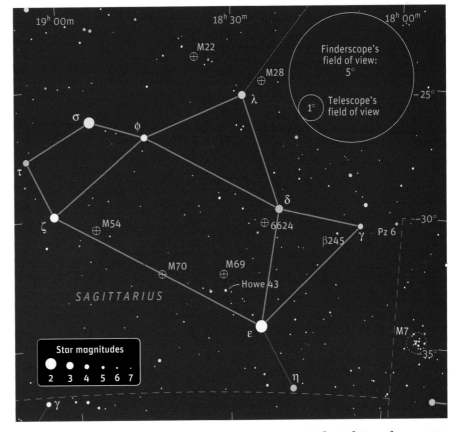

Skirting the Scutum Star Cloud

SCUTUM, THE SHIELD, is a modern constellation as constellations go; it was introduced by Johannes Hevelius in his 1687 atlas, *Uranographia*. Originally called Scutum Sobiescianum (Sobieski's Shield), it displayed the coat of arms of Hevelius's patron, King John III Sobieski of Poland. Scutum is the only constellation with political origins that is still in use today. (Among those that have gone by the boards are the Scepter of Brandenburg, Frederick's Glory, Charles's Oak, George's Harp, and the subject of some discussion on page 108, Poniatowski's Bull.)

Scutum is the fifth-smallest constellation in the sky, based on the official constellation boundaries established in 1930 by the International Astronomical Union, and it contains no star brighter than 4th magnitude. But its position along the hazy river of the Milky Way ensures us a region rich in deep-sky wonders.

On our evening constellation map on page 28, Scutum is drawn as a long, skinny diamond shape. Look for it about halfway between the "Facing South" label and Zenith, just behind the tail of Aquila, the Eagle, as he wings his way northeastward. On the constellation map you can also see that Scutum harbors a slightly brighter patch of the Milky Way. The Scutum Star Cloud is a glorious blizzard of tiny stellar jewels best appreciated with binoculars. It is highlighted on its northeastern edge by Scutum's grand star cluster, M11, my favorite of all open clusters.

M11 is easily visible in a finderscope as a small, misty patch of light sharing the field with 4th-magnitude Beta (β) Scuti 1.8° to the north-northwest. A 2.4-inch scope at low power shows M11 as a milky haze containing one 8th-magnitude star, probably in the foreground. High powers begin to resolve a handful of the cluster's brighter stars. My 4.1-inch refractor at 127× unveils a wonderfully rich cluster embracing a wealth of very faint stars. More than 100 glimmering points of light can be counted within a patch of sky just 13′ across. The stars appear to be gathered in patches and arcs, and a 9th-magnitude pair adorns the south-southeastern edge of the group.

In his 1844 observing guidebook, *A Cycle of Celestial Objects (Bedford Catalogue)*, the retired British admiral William H. Smyth wrote that M11 "somewhat resembles a flight of wild ducks in shape," and ever since then M11 has been known as the Wild Duck Cluster. Smyth's sketch, however, rounds off the apex of the supposedly V-shaped group so that it looks more like the head of a comet with an extra arc around the outside. As a result, observers have often been confused about where the skeins of ducks are supposed to be. A guide to recognizing Smyth's double **V** is shown at lower left.

A half degree northwest of M11 are two 6th-magnitude stars, both double. **Herschel VI 50,** also known as ADS 11719, is a very wide pair easily split at very low power. The 8th-magnitude companion star is 114″ south of the primary. **Σ2391** (Struve 2391) is less wide, with the 9.5-magnitude secondary 38″ north-northwest of the primary. At 30× both doubles are nicely split and fit in the same field of view.

This presents us with a good opportunity for color comparisons. Most observers see the primary star of the first double as golden or orange (it's a bright giant of spectral type $K1$), while the primary of Σ2391 is described variously as blue-white, white, or yellow-white (its spectral type is $A2$). All bets are off when it comes to the dimmer companions. Their measured color indexes are about 1.0 (yellow) and 0.5 (pale yellow), respectively, but subjective contrast effects often seem to intensi-

Above: The open cluster M11 is so dense and rich it could almost be taken for a globular cluster at first glance. But unlike a globular, it is as near as 6,100 light-years and as young as 200 million years. *Inset:* The "flight of wild ducks" pattern is clarified, based on W. H. Smyth's rough sketch accompanying his 1844 description of the cluster's appearance.

Dense fan

Long, flat V

fy or alter tints whenever two stars appear close together, especially when they differ greatly in brightness. Some have called Herschel VI 50 golden orange and bluish. What do you see?

A half degree northwest of our doubles we find the golden semiregular variable star **R Scuti.** It is the brightest of the class of variables known as the RV Tauri stars, pulsating yellow supergiants. R Scuti generally dips from about 5th to 6th magnitude over the course of a month or so, but at every fourth or fifth minimum it can drop to 8th magnitude. Judge its brightness by the magnitudes listed for the double stars in the table at right.

The huge dark-nebula complex **Barnard 111** lies just north of M11, our two double stars, and R Scuti. It stretches northward for about 2° and then takes a bend to the east. Barnard 119a, a degree-long crescent, lies just east of its southern part. Only binoculars or a very wide field telescope will encompass these large dark nebulae, but two smaller, darker regions within B111 are good targets for a narrower-field instrument.

Barnard 110 is a 10′ dark blotch 1° due north of Herschel VI 50 and ³⁄₄° due east of Beta Scuti. It's near an east-west pair of 8.5-magnitude stars; the western of these is yellowish, and the most conspicuous part of the nebula starts just to its north. B110 widens as it reaches north, where its borders grow less distinct.

Barnard 113 is a similar-size but more irregular dark nebula centered about ¹⁄₂° northeast of B110. The two are separated by a scattering of faint stars. Scutum is littered with dark nebulae, these two ranking among the easiest to pick out. While you're here, try for the equally dense but much smaller dark nebulae B107 and B106 a little west-southwest of

A very detailed map of the rich region around M11, for use at the telescope with a red flashlight. Start from 4th-magnitude Beta (β) Scuti, the top star of Scutum's narrow diamond shape on the constellation chart on page 28. From there you can't miss the glowing patch of M11 about 2° south-southeast. Using the map: North is up and east is left. To find which way is north in your eyepiece view, nudge your telescope slightly in the direction of Polaris; new sky enters the view from the north edge. Turn the map around to get properly oriented. If you're using a right-angle star diagonal you'll probably see a mirror image, which won't match correct-image maps. In this case remember to mentally flip the star patterns east-west, or take out the diagonal and view straight through to see a correct (but upside-down) image.

the star pair as seen on the map below.

These are just a few of the celestial riches lying along the boundaries of the beautiful Scutum Star Cloud. With its diminutive size, Scutum proves that sometimes good things come in small packages.

Some Sights Near the Scutum Star Cloud

Object	Type	Size/Sep.	Mag.	Dist. (l-y)	RA	Dec.
M11	Open cluster	0.2°	5.8	6,100	18ʰ 51ᵐ	–6° 16′
H VI 50	Double star	114″	6.2, 8.2	460, 250	18ʰ 49.7ᵐ	–5° 55′
Σ2391	Optical double	38″	6.5, 9.6	520, —	18ʰ 48.7ᵐ	–6° 00′
R Sct	Variable star	–	5–8	1,500	18ʰ 47.5ᵐ	–5° 42′
B111	Dark nebula	—	—	—	18ʰ 50ᵐ	–5°

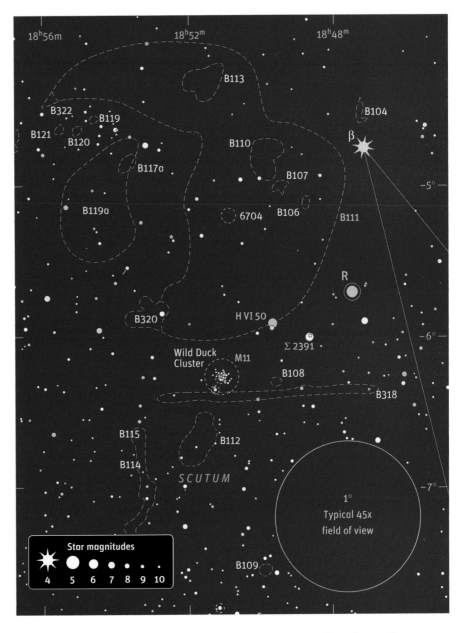

The Gem of the Milky Way

BEHIND THE TAIL of Aquila, the Eagle, a bright patch of Milky Way punctuates this otherwise hazy band. In his 1927 *Photographic Atlas of Selected Regions of the Milky Way*, Edward Emerson Barnard devoted a section entitled "The Great Star Cloud in Scutum" to this island of distant suns. He began, "This, the gem of the Milky Way, is the finest of the star clouds. It is interesting from many points of view. The main body is apparently made up of extremely minute stars. The great hammer-like head, however, looms up to the west with much coarser stars, as if it were much nearer to us."

Here we see the Sagittarius Arm of our Milky Way galaxy as it curves toward us. Countless stars along our line of sight create a dense concentration that is a wonder to behold in binoculars. The Scutum Star Cloud is bordered by the Great Rift, a dark dust lane that plunges from Cygnus to Scutum before broadening and turning west through Serpens Cauda toward

Altair

One of the Milky Way's brightest patches is the Scutum Star Cloud, about two fists at arm's length southwest of the 1st-magnitude star Altair. The image at right shows the star-cloud region at a much larger scale to match the finder chart.

Ophiuchus. This shadowy rift helps make the relatively bright star cloud one of the easiest features to see in the Milky Way.

The constellation where the star cloud resides is Scutum, the Shield. On our September all-sky chart on page 28, this is shown as a long, skinny diamond of 4th-magnitude stars. Scutum's brightest telescopic treasure is **M11,** a beautiful open cluster set against the northern border of the star cloud. In a finder M11 can be seen as a small, misty spot sharing the field with yellow Beta (β) Scuti.

Through a telescope M11 is a magnificent congeries of tiny stars. It is often called the Wild Duck Cluster for a V-shaped structure that I generally find less than evident at the eyepiece. Recently, however, when I observed M11 with the light dome of a nearby city washing out the faint stars, the comparison was a bit more apparent. With my 4.1-inch (105-millimeter) refractor at 153× the cluster appeared quite rich in faint stars, but a prominent dark lane in the shape of a very wide **V** sundered the bulk of the cluster from a straggling line of stars that did bring to mind a flock of migrating geese heading southeast. This left the rest of the stars gathered into a wedge-shaped group (another flock?) with the cluster's brightest star near the point. West of the bright star, which is probably in the foreground, the wedge was further broken by complex dark lanes plaited among the stars and reaching to the far edge of the cluster. (See page 120 for more about this star cluster.)

M11 was discovered in 1681 by Berlin astronomer Gottfried Kirch, who called it a nebulous star. A quote often attributed to Kirch, but actually from a 1715 paper by Edmond Halley, specifically mentions M11's foreground star: "It is of its self but a small obscure Spot, but has a Star that shines through it, which makes it the more luminous." In 1733 English clergyman William Derham laid claim to being the first to resolve the cluster's fainter stars, but he may have been referring to the Scutum Star Cloud as a whole.

Now let's visit **NGC 6664,** a more scattered group 20′ east of Alpha (α) Scuti. With 11 × 80 binoculars, Auke Slotegraaf of South Africa gives this charming description: "Beautiful! A soft glow, round, upon a murky field. A delicate object, surprisingly large, like percolated starlight." My less poetic notes, made while I was observing with a 6-inch reflector at 95×, simply log about 25 stars of 10th magnitude and fainter in an irregular group 16′ across. Nearby Alpha Scuti is yellow-orange.

NGC 6664's brightest star is usually **EV Scuti,** a Cepheid-type variable that ranges from magnitude 9.9 to 10.3 and back in a cycle of 3.1 days. At its dimmest,

EV Scuti has almost the same brightness as the cluster's other 10th-magnitude star 4′ to the northwest. Cluster Cepheids are treasured as important calibrators of the cosmic distance scale, and NGC 6664 is one of the relatively few clusters known to possess one.

In the 1960s and 1970s German astronomer Jörg Isserstedt compiled lists of 1,091 "prolate ellipsoidal stellar aggregates" once thought to be good tracers of our galaxy's spiral arms. **Isserstedt 68-603** is found in NGC 6664. It is visible, though unimposing, in my 6-inch reflector at 137× as a rather incomplete ring of 11th- to 13th-magnitude stars south of the cluster's center. The eastern arc is most obvious, while the northern edge is nearly nonexistent. The ring is oval east-northeast to west-southwest and about 6′ long. Most of these stellar rings are now believed to be chance arrangements of stars.

The open cluster **M26** is intermediate in concentration between M11 and NGC 6664. To track it down, scan 2.1° east from Alpha to 5th-magnitude Epsilon (ε) Scuti. Yellow Epsilon has a 7th-magnitude golden star just 6′ to the south. Put this pair in the western edge of a low-power field and drop 1.1° south to reach M26. Note also that a line from Alpha Scuti through 4.7-magnitude Delta (δ) Scuti points straight to M26. Delta, Epsilon, and M26 will all fit within a finderscope's field.

Although you may not be able to see M26 through your finder, you can use the triangle it makes with those two stars to pinpoint this impoverished cousin of M11. Indeed, in a casual sweep M26 might be mistaken for a mere asterism. My 4.1-inch refractor at 87× shows a pretty group of about 10 moderately faint stars within 8′ and many fainter stars like diamond dust. The brightest weighs in at magnitude 9.1 and sits at the southwest edge of the bunch.

The globular cluster **NGC 6712** lies 2.1° east-northeast of M26 and makes a nearly equilateral triangle with Alpha and Beta Scuti. My little refractor at 127× reveals a

5′ glow set amid a rich star field. A number of exceedingly faint stars at the very limits of vision are scattered over an unresolved blur, and one brighter star stands out in the northeast part of the halo.

Globular clusters have elongated orbits, and over eons they plunge through our galaxy's disk and then soar far out into its halo. The path of NGC 6712 carries it through the dense central regions of our galaxy, and this cluster is believed to have passed through the disk more times than most others have. One indication is that gravitational disruption has stripped the cluster of its lightest stars, which no doubt continue to move in orbits that take them into the Milky Way's vast halo. Many of the stars now in the halo may have been snatched from globular clusters.

The Scutum Star Cloud's Secrets

Object	Type	Mag.	Size	Dist. (l-y)	RA	Dec.
M11	Open cluster	5.8	13′	6,100	18h 51.1m	−6° 16′
NGC 6664	Open cluster	7.8	16′	3,800	18h 36.5m	−8° 11′
EV Sct	Variable star	9.9–10.3	—	3,800	18h 36.7m	−8° 11′
I68-603	Stellar ring	—	8′	3,800	18h 36.6m	−8° 17′
M26	Open cluster	8.0	14′	5,200	18h 45.2m	−9° 23′
NGC 6712	Globular cluster	8.1	10′	22,500	18h 53.1m	−8° 42′

Angular sizes are from various catalogs; most objects appear somewhat smaller in a telescope used visually.

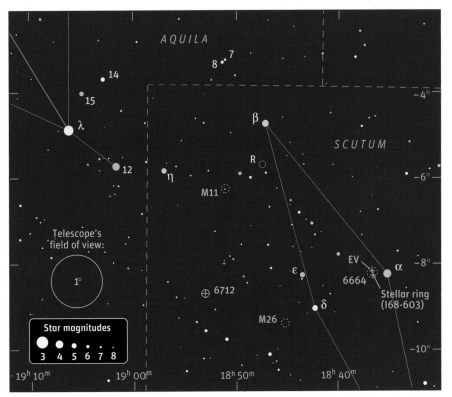

A small finderscope can easily encompass all the objects discussed in this essay (those whose names are in yellow). While exploring this region be sure to look in on R Scuti, the variable star lying only about 1° northwest of M11 (see page 121).

Lyre Lessons

AT THE CENTER of our all-sky map on page 28 you'll find **Vega** — the fifth-brightest star in the night sky and the brightest star in the constellation **Lyra,** the Lyre. According to one myth, the harplike lyre belonged to Orpheus, who charmed every living creature with its music. When his young wife, Eurydice, died, Orpheus went into Hades (the underworld) to seek her return. With the beauty of his music, he convinced Pluto, the king of the underworld, to let her go. Orpheus was warned not to look back at his bride until they had both reached the light of the upper world. When Orpheus stepped into the daylight he eagerly looked back, but it was too soon. Eurydice was still within the cavern, and as she vanished into the darkness he heard just a faint "Farewell." Thereafter Orpheus wandered heartbroken, and when he died the gods placed his lyre among the stars.

Lyra is the site of many deep-sky treasures suitable for a small scope, among them one of the most renowned multiple stars in the sky — **Epsilon (ε) Lyrae,** the Double-Double. Look for a 4th-magnitude star 1.7° east-northeast of Vega. Binoculars or a finder will reveal that this star is really a double. In fact, under good viewing conditions I can just split this pair with the unaided eye. Each of these stars is itself a double. Through a 3.6-

The Ring Nebula (M57) in Lyra appears as a small, dim, gray doughnut of light through most small telescopes, but a time-exposure photograph like this one from Akira Fujii reveals its true colors. The Ring is a planetary nebula, an aging star blowing much of its mass into surrounding space.

inch (92-millimeter) refractor, both star pairs are cleanly split at 94×, and 169× puts lots of space between them. The northern star (ε¹) consists of a 5.0-magnitude star with a 6.1-magnitude companion a very close 2.6″ to the north. The southern star (ε²) is a little tighter but more closely matched with a 5.3-magnitude star hugging a 5.4-magnitude companion 2.4″ to the east. All four stars appear white to slightly yellowish white. To split these close doubles you need good optics, atmospheric steadiness, and your scope must be nearly at equilibrium with the outdoor temperature.

Four 3rd- and 4th-magnitude stars southeast of Vega form a parallelogram that makes up the body of Orpheus's lyre. The nearest is the pair made up of **Zeta¹ (ζ¹),** a 4.3-magnitude white star, and **Zeta² (ζ²) Lyrae,** its 5.7-magnitude pale yellow-white companion, which lies 44″ away. They lie 1.9° southeast of Vega and are easily separated at 20×.

Next in the parallelogram we come to a more colorful double, **Delta¹ (δ¹)** and **Delta² (δ²) Lyrae,** 2° east-southeast of Zeta. At magnitude 4.2, Delta² is the brighter star and is distinctly reddish orange. Delta¹ is a prettily contrasting blue-white star at magnitude 5.6. This wide pair can be seen through steadily supported binoculars and is actually part of the sparse open cluster

Look for Lyra, the Lyre, directly overhead after sunset. It lies just west of Cygnus, the Swan. Lyra may appear to be an unassuming backdrop to zero-magnitude Vega but within its boundaries lie several beautiful pairs of stars, including Epsilon (ε) Lyrae, the famous "Double-Double." With low power you should be able to discern two stars that turn out to be double stars themselves.

Stephenson 1 (see the small chart below). A small telescope will show 10 more stars in the group (scattered mostly to the south of Delta[1] and Delta[2]), making this cluster appear 16' across.

Dropping 3.7° south-southwest from Delta, we come to a variable star whose brightness changes can be followed with the naked eye. **Beta (β) Lyrae** is an eclipsing binary whose components are so close that they are distorted into ellipsoids by their mutual gravitation and rapid rotation. The system shows a continuous change in brightness with the primary minimum coming once every 13 days. At maximum it is about as bright as 3.3-magnitude Gamma (γ) Lyrae, while at minimum Beta is about the brightness of 4.3-magnitude **Kappa (κ) Lyrae.** Compare Beta with these two stars and sooner or later you'll catch it in eclipse.

Our last parallelogram star is **Gamma (γ) Lyrae,** located 2° east-southeast of Beta. This bluish white star makes a nice binocular or low-power double with nearby orange **Lambda (λ) Lyrae.** Look ³/₅ of the way from Gamma to Beta to find the showpiece of Lyra: **M57**, the Ring Nebula. This planetary nebula is very small but distinctly nonstellar, even at 20×. Through the 3.6-inch refractor at 94×, M57 is a lovely little oval doughnut of light. The area within the ring appears brighter than the background sky. The star at the center of the Ring Nebula shows well on most photographs, but it is not visible through a small telescope and is a challenge even in large amateur instruments. The Ring Nebula is a cloud of gas being shed from the dying star at its heart. It is actually hourglass shaped but appears oval because we are seeing it nearly end on.

Our final target is the globular cluster **M56,** found 4° east-southeast of Lambda Lyrae with a 6th-magnitude orange star 26' to the northwest. Through a small scope at low power, M56 looks like a small, faint, round blur lying in a beautiful,

rich field of stars. It shows considerable brightening toward the center, and a 10th-magnitude star can be seen outside the western edge. When the air is clear a 6-inch scope at 200× will partly resolve this globular.

As you wander amid the sights of Lyra on late summer nights, let your imagination draw you into what John Milton described in *Il Penseroso* as the haunting melody that "drew iron tears down Pluto's cheek and made Hell grant what love did seek."

Lyra's Greatest Hits

Object	Type	Mag.	Dist. (l-y)	RA	Dec.
ε¹ Lyr	Double star	5.0, 6.1	162	18ʰ 44.3ᵐ	+39° 40'
ε² Lyr	Double star	5.3, 5.4	160	18ʰ 44.4ᵐ	+39° 37'
ζ¹ Lyr	Star	4.3	153	18ʰ 44.8ᵐ	+37° 36'
ζ² Lyr	Star	5.7	150	18ʰ 44.8ᵐ	+37° 35'
δ¹ Lyr	Star	5.6	1,000	18ʰ 53.8ᵐ	+36° 56'
δ² Lyr	Star	4.2	900	18ʰ 54.5ᵐ	+36° 54'
Stephenson 1	Open cluster	3.8	1,000	18ʰ 54.0ᵐ	+36° 52'
β Lyr	Variable star	3.3–4.3	900	18ʰ 50.1ᵐ	+33° 22'
γ Lyr	Star	3.2	634	18ʰ 58.9ᵐ	+32° 41'
κ Ly	Star	4.3	200	18ʰ 19.8ᵐ	+36° 04'
λ Lyr	Star	4.9	1,500	19ʰ 0.0ᵐ	+32° 09'
M57 (Ring Nebula)	Planetary nebula	9.7	2,000	18ʰ 53.6ᵐ	+33° 02'
M56	Globular cluster	8.2	31,000	19ʰ 16.6ᵐ	+30° 11'

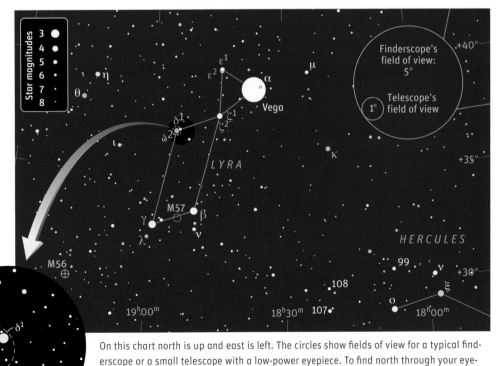

On this chart north is up and east is left. The circles show fields of view for a typical finderscope or a small telescope with a low-power eyepiece. To find north through your eyepiece, nudge your telescope toward Polaris; new sky enters the view from the north edge. (If you're using a right-angle star diagonal it probably gives a mirror image. Take it out to see an image matching the map.)

Exploring the Den of the Fox

HIGH OVERHEAD in the late-summer sky flies Cygnus, the Swan. You can find it near the center of our evening all-sky chart on page 28. (Discover what lies within the Swan starting on page 130.) Plotted there just south of Cygnus's head and neck is a small, dim, barely noticeable constellation: Vulpecula, the Little Fox. Only two of its stars barely made it onto the all-sky map (which shows stars as faint as magnitude 4.5). But this sly fox guards some famous telescopic treasures.

Vulpecula is one of nearly a dozen small constellations invented by the 17th-century Polish astronomer Johannes Hevelius; seven are still in use today. Hevelius originally named it Vulpecula cum Ansere (Fox with Goose) and drew on his star map a fox carrying a goose in its mouth. Since the goose is gone now, one can't help but wonder if the fox finally ate it. Maybe this is why Cygnus has taken flight.

Let's begin this small-scope observing project at a bright point that's easy to find: **Albireo,** which marks Cygnus's head. Albireo is the brightest star near the middle of the Summer Triangle of brilliant Vega, Deneb, and Altair; you should have no trouble spotting it without using any optical aid.

Albireo is one of the most beautiful telescopic double stars. A small instrument at very low power easily resolves it into a pair. The 3rd-magnitude bright component is golden yellow; the 5th-magnitude companion star appears clear blue-white. Their color contrast is very striking to most people, but a few have trouble seeing colors in stars. Low magnification helps; color contrasts look most vivid when stars are seen together. It may also help to put the stars slightly out of focus. The eye is not very sensitive to color in tiny,

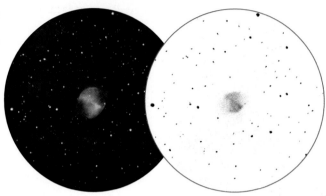

The Dumbbell or Apple Core Nebula (M27) is the most readily visible planetary nebula in the sky. I made the pencil sketch using a 4.1-inch refractor at 127×.

brilliant points, so spreading out the light just a little will make the tints more apparent.

We'll need to be a bit foxy to work our way from Albireo through the faint stars of Vulpecula, so let's stalk our deep-sky prey carefully. Use the detailed map at right, and make sure you know how to match what's on the map to what you see in your finderscope or main telescope's eyepiece. To do that you need to know the size and orientation of your instrument's view and whether it's a mirror image. If you're unsure about any of this, see the sections "Using a Star Chart" and "Navigating the Sky" on page 12.

The brightest star of Vulpecula lies 3.3° south of Albireo. This is Alpha (α) Vulpeculae, shining feebly at magnitude 4.4. It has no official common name, but in his classic book *Star Names, Their Lore and Meaning* (first published in 1899), R. H. Allen mentions two old dictionaries that define the word **Anser** as a star in Vulpecula. Alpha Vulpeculae would be in the right place, for on Hevelius's map it marks the fox's mouth holding the goose. Through a telescope you can see that Anser has an orange hue. A fainter orange companion, 8 Vulpeculae, lies 0.1° to its north-northeast.

Draw a line from Albireo through Anser and continue it onward for about 1½ times that distance. You'll arrive at a loose group of stars known as Brocchi's Cluster, Collinder 399, al-Ṣūfī's Cluster, or the **Coathanger.** The group is faintly visible to the naked eye; it was first described by the Persian astronomer Abu'l-Husayn al-Ṣūfī in AD 946 as a little cloud. In the 1920s it was mapped by Dalmiro F. Brocchi, chartmaker

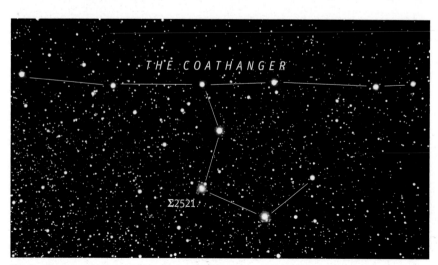

THE COATHANGER

Σ2521

The Coathanger asterism, 1.4° long, hangs upside down south of Albireo.

for the American Association of Variable Star Observers, who suggested that it be used to calibrate visual stellar photometers in the summer the way the Pleiades were used in winter. His name has stuck to the cluster ever since.

The Coathanger name refers to the shape formed by the group's 10 brightest stars. They are arranged in a pattern that resembles an old-fashioned coathanger with a wooden bar and a metal hook. Six stars form the east-west bar; the remaining four make the hook. You'll need to use a very low-power eyepiece, or perhaps even your finderscope or binoculars, to fit the Coathanger into your field of view; the bar is 1.4° long.

The Coathanger's hook contains a pretty quadruple star, **Σ2521** (Struve 2521). The 6th-magnitude primary star is orange. A 3-inch (76-millimeter) scope shows two 10th-magnitude companions to the northwest and east-northeast, widely split at low power. A 4-inch scope shows a fainter third companion (11th magnitude) closer in to the northeast; it's an easy split at 40×.

Although the Coathanger is often listed as a cluster, its stars are not physically related. The 10 brightest range from 220 to about 1,000 light-years away. Moreover, they are all moving in different directions. Only recently was the question of the Coathanger's status as a cluster put to rest, when data from the Hipparcos satellite provided a sufficiently accurate clue to each star's distance (*Sky & Telescope:* January 1998, page 65).

Our next target is the planetary nebula **M27** also known as the Dumbbell or Apple Core Nebula. It appears as a tiny glow in a good finderscope, noticeably unstarlike. You can pick it up by sweeping 3° due north from Gamma (γ) Sagittae, the point of the constellation Sagitta, the Arrow, as seen on our chart, or star-hop 2° southeast from 13 Vulpeculae.

The Dumbbell looks almost rectangular through a 2.4-inch scope at low power. With a slightly larger scope it begins to show its classic hourglass or apple-core shape. A 4-inch scope displays the nebula nicely, and you can begin to see faint extensions on each side that turn it into a larger football. Subtle variations in brightness become visible in the apple-core part of the nebula, including a narrow, brighter bar crossing it diagonally.

Our final object is more of a challenge, but two stars in Cygnus help locate it. Find the 4th-magnitude stars 39 and 41 Cygni on our chart at right. Draw a line from 39 through 41, continue on for the same distance, and you come to **NGC 6940** a large, dim open cluster. A 2.4-inch tele-

scope shows a handful of stars scattered over an elongated hazy background. A 4-inch reveals a few dozen sprinkled across an irregular, mottled glow about 1/2° across. Stars at the limits of vision seem to wink in and out of view, giving the cluster a look of fragile beauty.

Although Vulpecula reveals little to the unaided eye, the objects within make a fox hunt well worthwhile. Cygnus's head helped us with this month's deep-sky quarry. Turn to the next essay, and I'll show you where the Swan's tail leads us.

An Assortment of Foxy Delights

Object	Type	Mag.	Dist. (l-y)	RA	Dec.
Albireo	Double star	3.1, 5.1	380	19ʰ 30.7ᵐ	+27° 58′
Anser	Star	4.4	300	19ʰ 28.7ᵐ	+24° 40′
Coathanger	Asterism	5.2 to 7.2	220 to 1,000	19ʰ 26ᵐ	+20.0°
Σ2521	Multiple star	5.8, 10, 10, 11	450, –, –, –	19ʰ 26.5ᵐ	+19° 53′
M27	Planetary nebula	8	1,000	19ʰ 59.6ᵐ	+22° 43′
NGC 6940	Open cluster	6.3	2,600	20ʰ 34.6ᵐ	+28° 18′

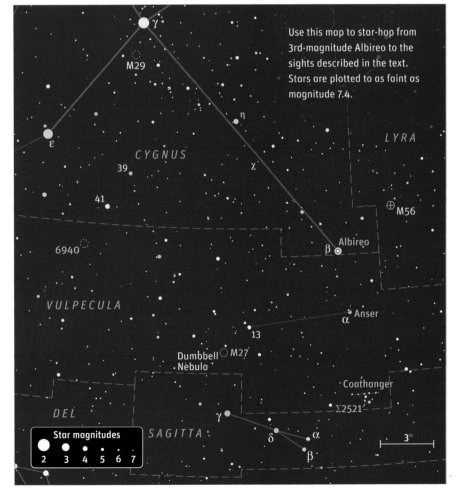

Use this map to star-hop from 3rd-magnitude Albireo to the sights described in the text. Stars are plotted to as faint as magnitude 7.4.

Chapter

4

At this time of year the northern sky holds a treasure-trove of deep-sky riches in Cepheus and Cassiopeia. High in the south and west lies the Milky Way with more fine sights in Cygnus, Sagitta, and other constellations. But don't ignore the rest of the heavens. There you'll find a bounty of double stars, a magnificent galaxy, and a beautiful planetary nebula.

Follow That Swan!

CYGNUS, THE SWAN, is nearly centered on the all-sky map on page 29. He is flying high overhead along the hazy band of the Milky Way with his wings outspread — his head marked by Albireo and his tail by the bright star Deneb. Let's shadow Cygnus and see what celestial sights he leaves in his wake.

We'll start well behind the Swan's tail with **M39,** the splashiest object in this essay's observing project. Aristotle may have been the first to record this star cluster in about 325 BC, describing it as cometary to his unaided eye. Messier 39 is a difficult naked-eye target but rather easy in binoculars. Sweep for it about three-fourths of the way from Deneb to the 4th-magnitude star Pi^2 (π^2) Cygni. A 3.6-inch (92-millimeter) telescope at 25× will show two dozen mixed bright and faint stars gathered in an equilateral triangle about the size of the full Moon. The cluster lies in a rich field of the fainter background stars of the Milky Way.

When you gaze at the hazy band of the Milky Way you are looking at a vast number of distant stars along the plane of our disk-shaped galaxy. It's a sight reserved for those who observe well away from outdoor lighting. The Milky Way is irregular — cleft and spotted with patches of darkness. These dark areas are not starless voids; they are clouds of interstellar dust — dark nebulae — blocking the light of the stars beyond.

Cygnus happens to contain the first dark nebula ever cataloged. **Le Gentil 3** is the third object listed in a scientific paper that was presented by the French astronomer Guillaume Le Gentil in 1749 and published in 1755. Le Gentil 3 can be seen with the naked eye or binoculars as a dark lane jutting across the Milky Way

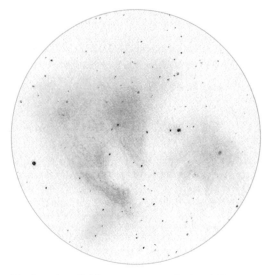

I used the lowest available power, 17×, to obtain a field of view 3¹/₂° wide for this negative drawing (dark stars, white sky) of the North America Nebula with my 4.1-inch Astro-Physics Traveler. A light-pollution filter (O III type) helped combat the skyglow at my observing site. The very faint nebulosity west (right) of "North America" is the Pelican Nebula.

between M39 and Deneb. Several other prominent dark nebulae can also be found in Cygnus, including a great, dark rift that runs along the southern side of the Swan's body and is easily visible to the unaided eye in dark skies. The end of this rift near Deneb is the Northern Coalsack.

Now let's move closer to Deneb, to the 7,000th object listed in J. L. E. Dreyer's famous *New General Catalogue* of 1888. Lying between Deneb and the 4th-magnitude star Xi (ξ) Cygni, this magnificent region of starbirth spans 2¹/₂° of sky, or five full-Moon diameters — but don't try to find it when the Moon is in the sky! **NGC 7000** (Caldwell 20) is faintly visible in binoculars under *very dark* conditions, but the best views come with a small telescope working at its lowest possible power and widest field of view.

I made the sketch above with a 4.1-inch f/6 refractor and an eyepiece giving only 17×. My drawing makes it obvious why NGC 7000 is called the North America Nebula. Resemblance to our continent's shape is striking, especially in the parts corresponding to Central America, the Gulf of Mexico, and Florida. This is also the brightest region of the nebula. Small telescopes with a focal length greater than about 750 mm cannot encompass the entire nebula in a single view. In that case you should concentrate on the Gulf

The North America Nebula is a popular camera target. Photographs like this one by Ben Gendre of Edmonton, Alberta, often bring out the nebula's vivid red color, which is never seen visually in a telescope of any size. Deneb is the bright star at right. Gendre used a lens of 150-mm focal length at f/4.5 for this 30-minute exposure on Kodak Pro PPF 400 film.

of Mexico. Don't expect the nebulosity to "jump out and grab you." It is quite subtle and takes careful study. Be sure you have spent at least 20 minutes in the dark before attempting to see the North America Nebula. This will give your eyes time to adapt and make the faint glow much easier to perceive.

61 Cygni is an interesting double star 5° south of Xi Cygni. It forms a nearly perfect parallelogram with the bright stars Deneb, Sadr, and Epsilon (ε). Both components of 61 Cygni are orange-dwarf stars, and their current separation of 31″ makes them an easy split for a small telescope at 30×. The brighter component is magnitude 5.2 and the companion is magnitude 6.0.

In 1792 the Sicilian astronomer Giuseppe Piazzi called special attention to 61 Cygni. Noting its unusual motion against the background stars, he dubbed it the Flying Star. In the 400 years since the invention of the telescope this star has "flown" more than a Moon's diameter northeast, toward Sigma (σ), as indicated by long arrows attached to the components in the *Millennium Star Atlas*. At 11.4 light-years 61 Cygni is the 14th-closest star system to our Sun and the 4th closest of those visible to the unaided eye. It was the first star to have its distance accurately measured (by the German astronomer Friedrich Wilhelm Bessel in 1838).

Almost 1.7° south of Sadr, Gamma (γ) Cygni, we find one of the least conspicuous star clusters in Charles Messier's catalog of deep-sky objects. In a 3.6-inch scope at 26×, **M29** can be seen as a little knot of six faint stars. It could easily be overlooked as a random clump in the star-rich area of the Milky Way. To me it looks like a set of back-to-back parentheses with three stars in each one. A few more stars may be glimpsed at high power, and scopes in the 4- to 6-inch range can up the count to 15 stars or so. But these added stars are as dim as those of the background Milky Way and do little to enhance M29's appearance as a cluster. Fellow astronomy-club members sometimes ask me to verify finds they feel unsure of at star parties. M29 is probably the most frequent example.

NGC 6826 is our final target, also known as the Blinking Planetary. It lies $2^1/_2$° southeast of the 4th-magnitude star Iota (ι) Cygni and just $1^1/_2$° east of the slightly dimmer star Theta (θ). In a 3.6-inch scope at low power NGC 6826 looks just like a

star, but at 50× it starts to reveal its true nature. A 4-inch scope will show NGC 6826, also called Caldwell 15, as a small, round, bluish green (planetlike) disk.

At 100× to 150× the nebula appears slightly oval and you can see its 10th-magnitude central star. Staring at this star with a 6-inch scope makes the nebula fade from view. But if you look a little off to one side of the star (use averted vision), the light from the nebula falls on a more sensitive area of the eye's retina and the planetary reappears. Alternating the two views produces a blinking effect, which gives this nebula its common name.

Celestial Sights in Cygnus

Object	Type	Mag.	Dist. (l-y)	RA	Dec.
M39	Open cluster	4.6	900	21^h 32.2^m	+48° 26′
Le Gentil 3	Dark nebula	—	—	21.2^h	+48°
NGC 7000	Bright nebula	—	500?	20.0^h	+44°
61 Cyg	Double star	5.2, 6.0	11.4	21^h 06.9^m	+38° 45′
M29	Open cluster	6.6	4,000	20^h 23.9^m	+38° 32′
NGC 6826	Planetary nebula	8.8	5,100	19^h 44.8^m	+50° 31′

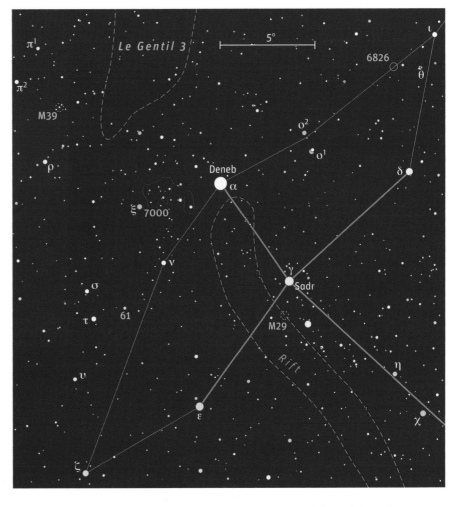

This close-up of Cygnus's tail section includes many stars not visible to the unaided eye. They are helpful when sweeping with a small telescope for the deep-sky objects described in this essay.

On the Wings of a Swan

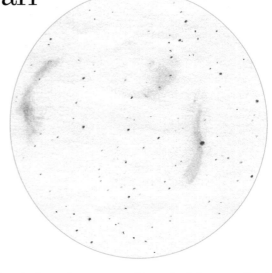

CYGNUS, THE SWAN, glides through the zenith on October's all-sky star chart on page 29. The leading edge of its outstretched wings are formed by Iota (ι), Delta (δ), Gamma (γ), Epsilon (ε), and Zeta (ζ) Cygni. Myriad deep-sky marvels, many within the grasp of small telescopes, nestle up to this distinctive line of stars.

We'll launch our tour with the planetary nebula **NGC 6826,** also known as Caldwell 15, near Iota in the Swan's northern wingtip. Iota shares the same finder field with 4th-magnitude Theta (θ) and 6th-magnitude **16 Cygni.** Centering 16 Cygni and viewing it with a low-power eyepiece will reveal a pretty pair of nearly matched yellow suns. Wait three minutes, and the slow westward march of the stars will bring NGC 6826 to the center of the field for you. The nebula will appear tiny, but nonstellar, at 50×. Boosting the magnification to 127× on my 4.1-inch (105-millimeter) refractor unveils a slightly oval robin's-egg-blue disk with a 10th-magnitude central star. Turn to the previous page to discover why NGC 6826 is nicknamed the Blinking Planetary.

Now let's move to the open cluster **NGC 6811,** located 1.8° northwest of Delta in the same finder field. My small refractor at 87× shows a 15′ swarm of about 40 faint to very faint stars in a nearly equilateral triangle. The cluster resembles an arrow with a fat head pointing west-southwest and a slender, slanted stalk. There's a paucity of stars in the center of the triangle.

Deep-Sky Wonders columnist Walter Scott Houston called for observations of NGC 6811 after a reader from Denmark likened it to a smoke ring of stars with a dark band running through its middle (*Sky & Telescope:* October 1986, page 426). The ensuing reports included

descriptions that compared it to the Liberty Bell, a butterfly, a frog, a clover (three-leaf and four-leaf), and even "Nefertiti's headpiece." Those using small scopes were the most apt to notice a dark interior enshrined by stars. What shape does this cluster suggest to you?

NGC 6866 also shares a finder field with Delta, which lies 3.5° west-northwest. The brightest star near this cluster is a 7th-magnitude, reddish orange star 24′ to the west. With my little scope at 87×, I see about 30 stars, faint to very faint, arrayed in a 10′ pattern that reminds me of a stunt kite. The kite is soaring to the northwest, and its southerly wingtip is bent. A 6′ knot of 17 stars forms the main body of the kite, the brighter ones gathered into a north-south bar.

The cute little cluster **NGC 6910** sits 33′ north-northeast of Gamma, the midpoint of the Swan's wings. At 87×, two yellowish stars of 7th magnitude and a pearly, split chain of eight 10th-magnitude stars unite in a Y-shaped pattern about 5′ long. A half

Encompassing the entire Veil Nebula, my drawing *(above)* was made October 5, 1997, using my 4.1-inch refractor and a 35-millimeter eyepiece (17×) with an O III filter. Compare my drawing to Bobby Middleton's image *(left),* composited from five hour-long exposures with a Takahashi 10-inch f/5 astrograph. Small scopes are good for tracing the overall shape and extent of large nebulae like the Veil, but the human eye has a problem detecting their fine structure.

dozen much dimmer stars join the scene.

NGC 6910 is embedded in an obscure section of **IC 1318,** the elaborate expanse of broken nebulosity surrounding Gamma Cygni. With a wide-field eyepiece giving 17× and a field of 3.6°, my little refractor reveals this to be an astounding complex. The three brightest patches lie 1.9° northwest, 0.8° east-northeast, and 1.1° east-southeast of Gamma. Each is aligned roughly northeast to southwest and appears about 30′ to 40′ long. To me the field stars seem to grow fainter toward Gamma, almost as if it were sitting at the bottom of a funnel.

Putting Gamma Cygni in the western part of a low-power field and scanning 1.8° south will bring us to **M29.** At 87×, I see a small, pretty gathering of six 9th-magnitude stars arranged in back-to-back parentheses of three stars each. Ten dimmer stars are sprinkled across the cluster. M29 has been described in some observing guides as a miniature dipper or a tiny Pleiades. Arizona observer Bill Ferris finds the resemblance so remarkable that he calls M29 the "Little Sisters." Color images of the group display a nice mix of blue and yellow stars.

We'll find our next target, **Ruprecht 173,** about three-quarters of the way from Gamma to Epsilon Cygni. Only low powers show this very large, coarse open cluster well. At 17×, I see 60 stars of 6th magnitude and fainter in 50′. Many of the brighter stars are arranged in a nearly cluster-spanning figure 8 with a fatter and dimmer southern half. A rich Milky Way star field skirts along and partly into the eastern edge of the group, where the variable star **X Cygni** resides. This pulsating yellow supergiant goes from around magnitude 5.9 to 6.9 and back in a period of 16.4 days. At its peak X Cygni is the cluster's brightest star.

Our final stop is the beautiful **Veil Nebula.** This supernova remnant is a lovely sight in a small telescope, but unless you observe under very dark skies you'll need an O III filter (one blocking most light except green) for a good view.

The brightest section of the Veil bears the designations NGC 6992 and NGC 6995, and together they are Caldwell 33. It is found a little southwest of a spot halfway between Epsilon and Zeta Cygni. This gently curving arc of gossamer light is more than 1° long and runs approximately north-south. The southern end widens and feathers out into ghostly tendrils reaching toward the west.

The Veil's other major part harbors the naked-eye star 52 Cygni, and for this reason many people prefer to start their visit to the Veil here. This section, called NGC

6960, or Caldwell 34, is a bit fainter than NGC 6992/5 but quite charming. It widens and forks to the south of 52 Cygni, while to the star's north it tapers to a point. If you center this northern tip in a low-power field, you can scan eastward to find NGC 6992/5.

A wide-angle eyepiece taking in both these large arcs offers perhaps the most engaging view. My sketch at 17× (opposite) shows a 3.6° field that nicely frames the entire Veil. The very faint wedge of nebulosity suspended between the northern ends of the brighter arcs is known as Simeis 229 or Pickering's Triangular Wisp. Harvard astronomer Edward C. Pickering mentioned this feature in 1906 after a long-exposure Veil photograph was examined by the observatory's first curator of astronomical plates, Williamina Fleming.

More Sights in Cygnus

Object	Type	Mag.	Size/Sep.	Dist. (l-y)	RA	Dec.
NGC 6826	Planetary nebula	8.8	27″ × 24″	5,100	19ʰ 44.8ᵐ	+50° 31′
16 Cygni	Double star	6.0, 6.2	40″	71	19ʰ 41.8ᵐ	+50° 32′
NGC 6811	Open cluster	6.8	12′	4,000	19ʰ 37.2ᵐ	+46° 22′
NGC 6866	Open cluster	7.6	10′	4,700	20ʰ 03.9ᵐ	+44° 10′
NGC 6910	Open cluster	7.4	7′	3,700	20ʰ 23.1ᵐ	+40° 47′
IC 1318	Emission nebula	—	4.0°	3,700	20ʰ 22ᵐ	+40.3°
M29	Open cluster	6.6	6′	3,700	20ʰ 24.0ᵐ	+38° 30′
Ruprecht 173	Open cluster	—	50′	4,000	20ʰ 41.8ᵐ	+35° 33′
X Cygni	Variable star	5.9–6.9	—	4,000	20ʰ 43.4ᵐ	+35° 35′
Veil Nebula	Supernova remnant	—	2.9°	-1,400	20ʰ 51ᵐ	+30.8°

Approximate angular sizes are from catalogs or photographs. Most objects appear somewhat smaller in a telescope used visually.

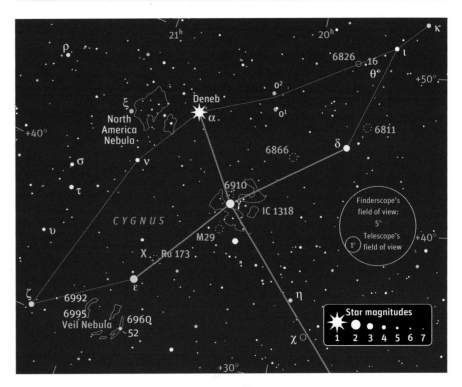

The Arrow and the Dolphin

ON OCTOBER EVENINGS Delphinus, the Dolphin, appears to be leaping out of the "watery" constellations to the south and east (see page 29). Nearby Sagitta, the Arrow, seems poised for a flight over the Dolphin's head. While diminutive in size, Delphinus and Sagitta are in an area of the sky rife with deep-sky splendors.

We'll begin our tour in Sagitta with **M71,** which long suffered an identity crisis. Some astronomers maintained that M71 was an open star cluster like the Beehive or the Pleiades, but richer. Others contended that it was a globular cluster like M13 in Hercules, only looser. A recent meeting of the minds has placed it firmly in the latter category — much to the relief of the cluster, no doubt.

M71 poses prettily between Delta (δ) and Gamma (γ) Sagittae, two 4th-magnitude orange stars in the body of the arrow. Look for it about halfway between the stars and just a little south of a line connecting them. In 14 × 70 binoculars, M71 is a small hazy patch that I'd call moderately bright and slightly irregular in shape. With my 4.1-inch refractor at 28×, I see an obvious misty spot that is brighter in the center and shows hints of very faint stars. Three stars sparkle at or near the edge: one north, another east, and the third (brightest) toward south. At 153×, many faint stars burst into view, strewn across an unresolved, cottony background.

At low power, you may have another star cluster in the field with M71 and yet fail to notice it. The scanty open cluster **Harvard 20** (H20) lies just 28′ to the south-

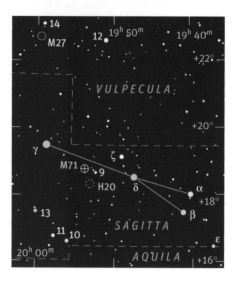

Sagitta's two clusters, M71 and H20, lie less than 5° south of M27, the Dumbbell Nebula in Vulpecula (see page 127).

southwest. Look for two 9th-magnitude stars, the brightest in the area. At 87×, I see the inconspicuous group engulfing the eastern member of the pair. A star count at 127× gives me 11 possible members plus several stragglers to the east-southeast. The enclosed 9th-magnitude star sits a little west of center. Although H20 does not stand out well against the backdrop of the Milky Way, it is a true cluster with about 100 known members and an age of 30 million years. The 9th-magnitude star near the cluster's western edge is a

Above: The tiny constellations Sagitta and Delphinus are easy-to-spot patterns of mostly 4th-magnitude stars nestled north (above) and east (left) of the bright star Altair. This region is home to the intriguing star groups and clusters described above, along with the distinctive Coathanger asterism at top right (see page 126). *Right:* The globular cluster M71 (top left) and loose open cluster H20 (bottom right) show well in this photograph by Martin C. Germano at Mount Pinos, California.

foreground object, but the embedded one is a yellow supergiant believed to be a member of the group.

Now let's move to the nose of the Dolphin, where we'll find two nice double stars. The golden star marking the Dolphin's snout is 4th-magnitude **Gamma (γ) Delphini.** Through a telescope, a 7th-magnitude star of similar hue lies 15′ south-southeast, sharing a low-power field. My little refractor at 28× reveals both stars as doubles — but they are still quite close! At 47× you'll get a nicer view, but if the air is unsteady don't be surprised if you need to double that magnification. Gamma consists of a 4.4-magnitude yellow-orange primary and a nicely contrasting, 5.0-magnitude yellow-white companion 9″ to the west. The dimmer pair, **Σ2725,** is a 7.5-magnitude star with an 8.2-magnitude secondary 6″ to the north, both yellow-orange.

A very pretty triple star can be seen near Beta (β) Delphini, the star that marks the opposite corner of the Dolphin's diamond-shaped head. **Σ2703** lies a mere 13′ to the northwest, its roughly 8th-magnitude components being arranged in a skinny triangle. They appear well separated at 28×. The easternmost member is a white star, accompanied by a yellow-orange star to the west-northwest and a slightly fainter yellow star about three times as distant to the southwest.

Although it inhabits a region of the sky devoid of bright landmarks, I can't resist directing you to my favorite asterism, the **Toadstool.** It lies nearly due east of the Dolphin's nose by exactly 5°. In fact, the simplest way to find it would be to put Gamma Delphini in the southern part of a low-power field, turn off your telescope's drive if it has one, and come back for a look in 20 minutes. The Toadstool (which we'll return to at the end of this article) should then be drifting toward the center of your field of view.

For those who prefer to skip the wait, try slowly scanning eastward with your lowest-power eyepiece. It is easier to arrive at your goal if you know how much of the sky your eyepiece shows. Eyepiece manufacturers often state the *apparent* field of an eyepiece, but you need to know its *true* field. For more on this, see page 13.

While counting star fields, you may notice a small dim smudge after sweeping 3.6° eastward — the faint globular cluster **NGC 7006** (Caldwell 42). Pausing for a look at 68× simply lets me see a round, faint glow with a slightly brighter core. This globular journeys into the far reaches of the Milky Way with a highly eccentric orbit that takes it from 60,000 to 330,000 light-years from the galactic center.

Dropping back to low power, position NGC 7006 a little south of center and continue east for another

1.4°. This should bring you to the Toadstool, an asterism of 9th- through 12th-magnitude stars about 13′ across. My little scope at 68× shows the celestial mushroom's stem formed by five stars in a **V** shape opening toward the northeast. The top of the eight-star cap faces southwest. The Toadstool's shape shows very well in the *Millennium Star Atlas,* which plots 11 of the 13 stars visible through my little refractor. Deep-sky writer Philip S. Harrington sees these stars grouped differently; he's dubbed them Dolphin's Diamonds.

While asterisms are generally chance alignments of stars with no astrophysical significance, I find their eye-catching shapes enchanting. Do you have a favorite asterism?

Star Groups Thick and Thin

Object	Type	Mag.	Size/Sep.	Dist. (l-y)	RA	Dec.
M71	Globular cluster	8.2	7′	13,000	19ʰ 53.8ᵐ	+18° 47′
H20	Open cluster	6.8	6′	5,000	19ʰ 53.2ᵐ	+18° 20′
γ Del	Double star	4.4, 5.0	9″	101	20ʰ 46.7ᵐ	+16° 07′
Σ2725	Double star	7.5, 8.2	6″	125	20ʰ 46.2ᵐ	+15° 54′
Σ2703	Triple star	8.4, 8.4, 8.7	25″, 76″	700	20ʰ 36.8ᵐ	+14° 44′
Toadstool	Asterism	9 to 12	13′	—	21ʰ 07.4ᵐ	+16° 18′
NGC 7006	Globular cluster	10.6	3′	140,000	21ʰ 01.5ᵐ	+16° 11′

Most sizes or separations are catalog values. Most objects appear somewhat smaller in a telescope used visually.

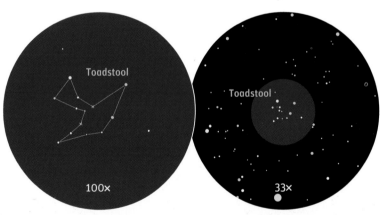

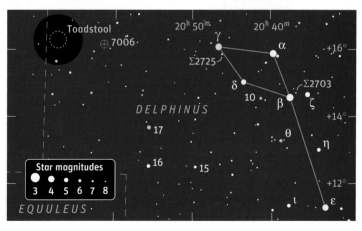

The Toadstool asterism in Delphinus is not easy to make out in binoculars. It is best studied in small scopes at low to medium power. The eyepiece views above the chart include stars to magnitude 11 from the *Millennium Star Atlas* along with two fainter stars, marked ×, that complete the figure. The mushroom is standing on its cap in these north-up views. The new *Uranometria 2000.0* atlas labels it French 1.

Messier Hunting with the Horse and the Water Boy

OBSERVING ALL the Messier objects is a goal pursued by many amateur astronomers and a fine way to hone your observing skills. A good 3-inch (75-millimeter) telescope under moderately dark skies is sufficient to bag them all. Here we'll ferret out the Messier objects of Pegasus and Aquarius. While two objects are easy targets for a small telescope, the other two may prove to be a challenge. Let's start with the constellation Pegasus, which contains the brightest and easiest to find of the four Messiers.

The 18th-century astronomer Jean-Dominique Maraldi's interest in de Chéseaux's Comet (which appeared in August 1746) led him to discover two of this month's globular clusters, M15 and M2. On September 7, 1746, he observed M15 as "a fairly bright, nebulous star which is composed of many stars," a somewhat perplexing description.

Look for **M15** (the Great Pegasus Cluster) just off the nose of Pegasus, the Flying Horse. This cluster may have the densest core of any globular in our galaxy. It is theorized that M15's stars fell toward its center within the first few million years of the group's 12-billion-year lifetime. Although no globular cluster is unam-

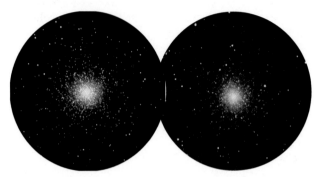

Look in Pegasus for M15 *(left)* and about a fist-width at arm's length below it for M2 *(right)* in Aquarius. These globulars rival each other in brightness, though M2 is a little dimmer and farther away from us.

biguously known to contain a central black hole, M15 is considered a good candidate, possibly containing one with 2,000 solar masses. Alternatively, the core density may simply be due to a population of stellar remnants such as neutron stars and white dwarfs.

M15 is also the first globular discovered to contain a planetary nebula — Pease 1. It has been observed in telescopes as small as 10 inches but is considered a

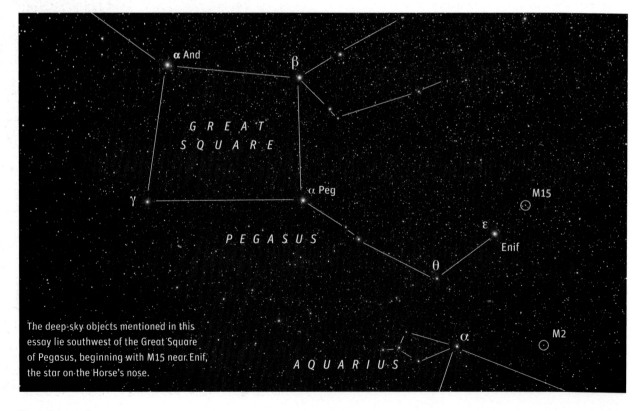

The deep-sky objects mentioned in this essay lie southwest of the Great Square of Pegasus, beginning with M15 near Enif, the star on the Horse's nose.

tough visual target even in large amateur instruments.

To find M15, draw a line from Theta (θ) to Epsilon (ε) Pegasi and continue for a little more than half that distance again. Golden-hued Epsilon is also known as Enif and it marks the horse's nose. At magnitude 6.4, M15 has been spotted with the unaided eye under very dark skies by some keen-sighted observers. Through a small finder it resembles an out-of-focus star. My 14 × 70 binoculars show a fairly bright, round, hazy patch surrounding a very small, bright nucleus. Through my 4.1-inch refractor at 47×, the halo of the cluster looks mottled. At around 200× the cluster appears slightly oval and some of the outer stars pop into view, but the center remains unresolved.

Our next globular cluster, **M2,** lies 13° south of M15 and makes a nearly right triangle with Alpha (α) and Beta (β) Aquarii. M2 is a near twin to M15 in size, brightness, and distance (38,000 light-years), but it is less condensed. At 20× a 2.8-inch scope shows a bright, round patch of light surrounding an elusive, starlike nucleus. In the 14 × 70 binos, the bright nucleus is more obvious and nearly stellar. Magnified at 200×, the cluster appears distinctly oval and sparkles with many very faint stars.

Two dim Messier objects, M72 and M73, are located in southwestern Aquarius, and if you'd like to observe the entire Messier catalog, you'll need to tackle these more difficult targets.

M72 is located 3.3° south-southeast of 3.8-magnitude Epsilon (ε) Aquarii and is the faintest globular cluster on Messier's list at magnitude 9.3. An impressive 55,000 light-years away, the stars of M72 are very difficult to resolve in a small telescope. My 70-mm binoculars shows a very small and dim fuzzy patch. A 9.4-magnitude star lies 5′ to the east-southeast. At 127× in my 4.1-inch refractor the cluster appears grainy, but there is no true resolution into stars.

Our last Messier object is **M73,** found 1.3° east of M72. Don't blink or you'll miss it! M73 is usually classified as an asterism of unrelated stars, but some believe it might be a multiple-star system or the remains of a dispersed cluster. M73 consists of four stars at magnitudes 10.3, 11.1, 11.9, and 11.9 in a Y-shaped group a mere 1′ across. Use magnifications of around 100× to darken the sky background and help bring out the fainter stars.

If you don't find M73 sufficiently challenging, try for the nearby planetary nebula **NGC 7009,** often called the Saturn Nebula. You can star-hop from M73 using an eyepiece that gives about a 1° field of view. Put M73 at the western edge of the field and you will see a 7.1-magnitude star on the opposite side. Place this star in the southern part of your field and there will be a 7.0-magnitude star near the northern edge. Position that star near the southwestern edge of the field and 8.3-

On this chart north is up and east is left. The circles show fields of view for a typical finderscope or a small telescope with a low-power eyepiece. To find north through your eyepiece, nudge your telescope toward Polaris; new sky enters the view from the north edge.

magnitude NGC 7009 will be the brightest object in the northeast. At 29× in my 4.1-inch scope the planetary looks stellar, and at 153× it appears small, oval, and bluish gray. The faint extensions or ansae that give NGC 7009 (Caldwell 55) its Saturn-like appearance are not easily visible in scopes less than 10 inches in aperture.

Star Quest 2007 10/12/07
M72 Very challenging. No resolvable stars even in Teeter 17
NGC 7009

Attractions in Pegasus and Aquarius

Object	Type	Mag.	Dist. (l-y)	RA	Dec.
M15	Globular cluster	6.2	34,000	21ʰ 30.0ᵐ	+12° 10′
M2	Globular cluster	6.5	38,000	21ʰ 33.5ᵐ	– 0° 49′
M72	Globular cluster	9.3	55,000	20ʰ 53.5ᵐ	–12° 32′
M73	Asterism	9.0	—	21ʰ 00.0ᵐ	–12° 38′
NGC 7009	Planetary nebula	8.3	3,000	21ʰ 04.0ᵐ	–11° 22′

A Fish Tail to Get Your Goat

THE CONSTELLATION CAPRICORNUS is due south on this month's all-sky star chart on page 29. Capricornus is commonly called the Sea Goat, though its name literally means Goat Horn. In Greco-Roman mythology, the Sea Goat is often associated with the god Pan. It is said that the Olympian gods were feasting along the banks of the Nile when Pan warned them that the dreadful monster Typhon was approaching. The gods quickly disguised themselves as animals, but in a panic (a word derived from his name) Pan leapt into the river — where the part of his body below water became a fish and the portion above became a goat.

This story was apparently concocted to explain a creature of much more ancient origin from the Sumerians perhaps five or six millenia ago. Their Lord of the Earth, Enki, came to be symbolized by a goat-fish or ram-fish, which worked its way into the sky sometime during the first millennium BC. When the Greeks systematized the signs of the zodiac more than 2,000 years ago, the Sea Goat was indeed a fitting symbol for this sky region. At that time the Sun was at the winter solstice, its lowest point in the sky, when in Capricornus. What better animal than a goat for climbing out of winter's depths! The latitude on Earth over which the Sun hovers on the December solstice is still called

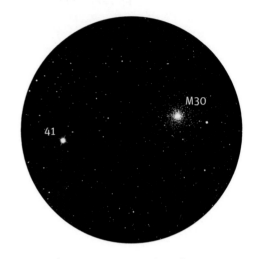

the Tropic of Capricorn, even though precession long ago moved the solstice into Sagittarius.

The head of the Sea Goat is marked by **Alpha (α) Capricorni,** also known as Algedi. It is composed of two stars that are far enough apart (381″, slightly more than 0.1°) to split with the unaided eye. They are distinctly colored in binoculars or a small telescope. Through my 4.1-inch (105-millimeter) refractor at 17×, 4.3-magnitude α^1 appears yellow-orange, while I see 3.6-magnitude α^2 as golden yellow (slightly less orange). This is an optical pair of unrelated stars; the brighter one is several hundred light-years closer to us than its seeming companion.

Each saffron gem has a faint companion of its own, also visible at 17×. The attendant to α^1 is 46″ to its southwest, and that of α^2 is a spacious 158″ to its south-south-east. To me they look to be 9th or 10th magnitude.

Many other fine multiple stars adorn Capricornus. South-south-east of Algedi by 2.3° we find 3rd-magnitude **Beta (β) Capricorni,** or Dabih. Beta Cap is a very wide, very pretty triple, well split through my little scope at 17×. The 3rd-magnitude yellow primary has

Above: The pretty little globular cluster M30 is an easy find, situated just 0.4° west-northwest of 5.2-magnitude 41 Capricorni. *Left:* The constellation Capricornus as imaged by Akira Fujii. Compare it with the map at right.

a 6th-magnitude bluish secondary 205″ to the west and a 9th-magnitude tertiary 227″ to the southeast.

From Beta Cap, drop 4.3° south to 5th-magnitude **Sigma (σ) Capricorni.** The two should fit together in the field of most small finders. This is a lovely double through my scope at 17×. The primary appears orange, and the 9th-magnitude secondary 55″ to its south looks distinctly reddish orange despite its faintness.

A little triangle of double stars lies 2½° east-north-east of Sigma. The northernmost of the three is **Rho (ρ) Capricorni.** At 17× I see a 5th-magnitude yellow-white star with a 7th-magnitude yellow-orange companion a whopping 256″ to the south-southeast.

A closer pair is **Omicron (o) Capricorni,** the southern-most of the three. The color contrast is subtle, but to me the 6th-magnitude primary looks white and the 7th-magnitude secondary 21″ to the southwest appears slightly yellowish white. Although this is the closest double we have visited so far, it is still nicely split at 17×.

Our remaining star of the trio is **Pi (π) Capricorni.** The color of the 5th-magnitude primary is not obvious, but when I compare it to the white primary of Omicron, I see Pi as slightly bluish. I used 118× to isolate the 8.5-magnitude companion lying just 3.4″ south-southeast. This is a challenging double for a small scope from my upstate New York latitude of 43°. Capricornus rides low in my sky, where poor atmospheric seeing often blurs the view.

Delta (δ) Capricorni is the brightest star of the Sea Goat and can help in finding our final target, the globular cluster **M30.** From Delta, star-hop 3° south-southwest to 4.7-magnitude Kappa (κ) Capricorni. Then drop 4.4° south to 5.2-magnitude 41 Capricorni. You will find M30 just 0.4° west-northwest of 41 Cap, comfortably sharing the same low-power field (see image at upper left).

I can easily spot M30 as a small fuzzball in my 4.1-inch refractor at 17×. Boosting the magnification to 87× reveals considerable brightening toward the center and seven faint stars in the outer halo. The halo stars stand out only with averted vision (the deep-sky observer's trick of looking slightly off to the side of an object so its light will fall on a more sensitive area of the eye's retina). The entire cluster appears about 5′ across in my scope, though it is cataloged as being more than twice as large. A 9th-magnitude field star lies outside its western edge.

Those who live farther south can see M30 higher in the sky and might pick out details that I usually need a larger telescope to see. At high powers this globular appears quite asymmetric, with the brightest halo stars arranged in four stubby arms. Two form a **V** shape radiating from the cluster's northern edge. The other two emerge from its southeastern and southwestern sides and extend northward. Comically, this shape brings to mind a fat, round body with legs slightly spread and arms held out at an angle from the sides.

On some fine late-summer or autumn night, try the wonders of Capricornus with your small scope. They won't get your goat.

M30 StarQuest 10/13/2007
IN New Burgess 8mm - won @ raffle today, nice sprinkle of
stars dropping down from cluster. Not obvious in "opposite" photo

Selected Sights in Capricornus

Object	Type	Size/Sep.	Mag.	Dist. (l-y)	RA	Dec.
α² – α¹ Cap	Optical double	381″	3.6, 4.3	108, 700	20ʰ 18.0ᵐ	–12° 33′
α²	Double star	158″	3.6, 10	108	20ʰ 18.0ᵐ	–12° 33′
α¹	Double star	46″	4.3, 9	700	20ʰ 17.6ᵐ	–12° 30′
β Cap	Optical triple	205″, 227″	3.1, 6.1, 9.1	340, 315, 470	20ʰ 21.1ᵐ	–14° 47′
σ Cap	Optical double	55″	5.3, 9.2	700, —	20ʰ 19.4ᵐ	–19° 07′
ρ Cap	Optical double	256″	4.8, 6.6	100, 500	20ʰ 28.9ᵐ	–17° 49′
o Cap	Optical double	21″	5.9, 6.7	240, 140	20ʰ 29.9ᵐ	–18° 35′
π Cap	Double star	3.4″	5.2, 8.5	660	20ʰ 27.3ᵐ	–18° 13′
M30	Globular cluster	11′	7.5	27,000	21ʰ 40.4ᵐ	–23° 11′

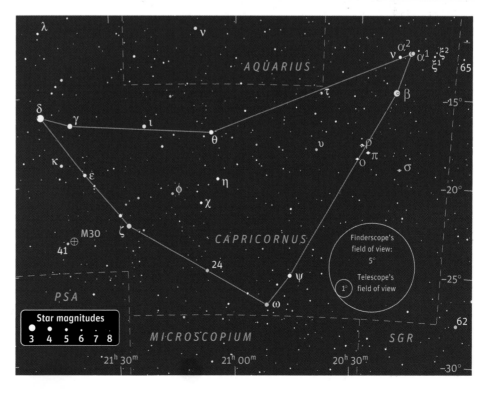

Leaping Lizards

LACERTA, THE LIZARD, is a modern constellation introduced by Johannes Hevelius in his 1687 atlas, *Uranographia*. He also referred to Lacerta as Stellio, the Stellion — a coastal Mediterranean newt with starlike markings on its back. Lacerta zigzags across the Milky Way above Pegasus, the Winged Horse, and just northeast of Cygnus, the Swan. Our celestial lizard is an inconspicuous constellation whose brightest star shines at magnitude 3.8, but from a dark location it adorns the sky with a pretty sprinkling of faint stars scattered across the haze between the brighter Milky Way constellations of Cygnus and Cassiopeia. Only two stars in Lacerta have Bayer Greek-letter designations: white, 3.8-magnitude Alpha (α) Lacertae, and yellow-orange, 4.4-magnitude Beta (β) Lacertae. Our first celestial target is the open cluster **NGC 7243,** lying approximately 2.6° west of Alpha and making an isosceles triangle with Alpha and Beta.

This book often highlights objects (prefaced by the letter M) from Charles Messier's catalog, which includes some of the brightest and best treasures of the deep sky. But there are many other small-scope wonders to be found, particularly in the *New General Catalogue of Nebulae and Clusters of Stars (NGC)* compiled by John Dreyer in 1888. Although many of the 7,840 NGC entries are beyond the reach of a small telescope or binoculars, NGC 7243 is one object that is approachable through both. In fact, it is included in the Astronomical League Deep Sky Binocular Club's list of 60

The open clusters NGC 7209 *(above)* and NGC 7243 *(facing page)* make good targets for observers with binoculars and small telescopes. Look for them just slightly to the west of Lacerta. North is up in both images.

non-Messier objects observable with 7 × 50 binoculars. It is also number C16 in the Caldwell list, compiled by British astronomer and astronomy popularizer Patrick Caldwell-Moore.

In my 8 × 50 finder I can pick out several very faint stars in NGC 7243 over a hazy background. Once, while observing this group through 14 × 70 binoculars, I nicknamed it the Broken Heart. To me, its 14 brightest stars roughly outline a heart symbol broken open along one side where one of the stars seems greatly misplaced. All likeness to a heart shape is lost through my 4.1-inch refractor, however, which shows about 50 moderately bright to faint stars within 20′ at 68×. The stars are distributed in clumps surrounding a central isosceles triangle of one 10th- and two 9th-magnitude stars. The southeastern corner of the triangle is the double star **Σ2890.** Its 9.3-magnitude primary cuddles up to a 9.7-magnitude companion 9″ to the northnortheast.

A second cluster from the League's Binocular Club resides in Lacerta. **NGC 7209** can be found 3.8° southsouthwest of NGC 7243 and 2.7° due west of the 4.6-

Lacerta, the Lizard, appears to be scuttling up from Pegasus to Cepheus. Look for three open clusters lying between the Lizard's zigzag-shaped body and one of the outstretched wings of Cygnus, the Swan, to its west.

magnitude star 2 Lacertae. NGC 7209 appears nebulous through my 8 × 50 finder, with only one or two faint stars visible within, but through 14 × 70 binoculars it appears vaguely rectangular and very rich in faint stars. The cluster sports two concentrations of slightly brighter stars. With such a faint cluster, the star count varies considerably with observing conditions. My 4.1-inch scope at 68× reveals between 40 and 50 moderately faint to very faint stars within a 23'-wide patch of sky. The brightest stars snake north to south across the group in a wide S-shape. A 6.2-magnitude orange-yellow star lies just outside the northern border of the cluster. Three stars of magnitudes 8.5, 8.5, and 8.8 cradle the cluster's edges to the west-north-west, southwest, and south-southeast, respectively. The dimmest of the three has an orange hue.

NGC 7209 contains the ex-eclipsing binary **SS Lacertae.** This variable showed a dip in brightness when one of its stars passed in front of the other. Studies indicate that SS Lacertae's eclipses began in the late 1800s and ceased sometime near the middle of the 20th century. This bizarre behavior has been attributed to an unseen third star in the system whose gravitational perturbations affect the dynamics of the system. SS Lacertae is visible as a 10.1-magnitude white star in the southwestern quadrant of NGC 7209. The intricate dance of these three suns will not produce eclipses for Earthly observers until at least sometime in the 4th millennium.

Just 1.4° south of NGC 7209 is an attractive curve of stars on which to try your color perception. Their magnitudes from north to south are 6.2, 6.5, and 5.1. Through my 4.1-inch refractor I see them as white, bluish, and yellow-orange. What colors do you perceive?

The *New General Catalogue* was followed by two supplementary *Index Catalogues (IC)* in 1895 and 1908. Together they list an additional 5,386 objects. Many of the IC objects are quite faint, but the open cluster **IC 1434** in Lacerta is within the reach of a small telescope. IC 1434 is located 2.1° west-northwest of Beta

On this chart north is up and east is left. The circles show fields of view for a typical finderscope (5°) or a small telescope with a low-power eyepiece (1°). To find north through your eyepiece, nudge your telescope toward Polaris; new sky enters the view from the north edge. (If you're using a right-angle star diagonal it probably gives a mirror image. Take it out to see an image that matches the map.)

Lacertae. You can find it by drawing a line from orangish 4.3-magnitude 5 Lacertae through 4.6-magnitude 4 Lacertae and continuing for exactly twice that distance again. Most of the stars in IC 1434 are extremely dim, but using high magnifications will darken the sky background and help unveil their feeble light. My little scope at 17× shows a faint, fuzzy patch with four dim stars. At 87×, I see five moderately bright and several extremely faint stars over a misty background about 7' across. Four of the brighter stars are arranged in a simple kite shape with the brightest of the five to their southeast. A power of 153× brings out many barely perceptible stars glittering over a patchy haze.

Although nearly barren to the naked eye, the Lizard harbors deep-sky treasures well worth seeking. Next we'll go looking for more deep-sky wonders in the constellation immediately to the east of Lacerta.

Pretty Sights in a Faint Constellation

Object	Type	Mag.	Dist. (l-y)	RA	Dec.
NGC 7243	Open cluster	6.5	2,800	22ʰ 15.3ᵐ	+49° 53'
Σ2890	Double star	9.3, 9.7	2,800	22ʰ 15.2ᵐ	+49° 52'
NGC 7209	Open cluster	6.7	3,000	22ʰ 05.2ᵐ	+46° 30'
SS Lacertae	Binary star	10.1	3,000	22ʰ 04.7ᵐ	+46° 26'
IC 1434	Open cluster	9.0	—	22ʰ 10.6ᵐ	+52° 50'

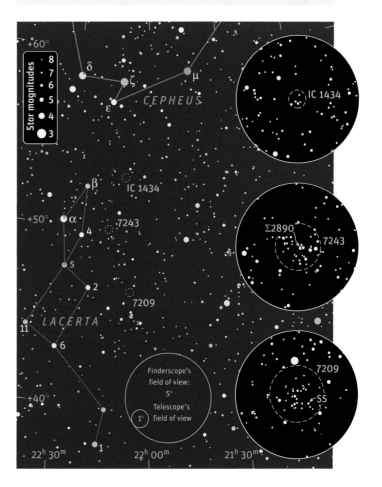

Arcing Through the Autumn Sky

THE CRISP NIGHTS of November offer many of us in the Northern Hemisphere a threefold blessing: early darkness, bug-free observing, and frequently transparent skies. They also harbor a panoply of deep-sky wonders for anyone with a small telescope.

This time we'll seek a group of objects that gather around zenith for observers in north temperate climes. Looking at our all-sky chart (page 30), just above the *Z* in the label *Zenith,* you'll see a curve of three 4th-magnitude stars belonging to the constellation Andromeda. In the night sky you can locate this group of stars by noting its position above a distinctive and very large star pattern — the Great Square of Pegasus (see page 136 for a photograph of this asterism).

The southernmost star in the curve is Iota (ι) Andromedae, which will serve as the point of departure for our journey. Some 2° west and a little south of Iota, you'll see the 6th-magnitude star 13 Andromedae — the brightest star in the area. Our first target, the tiny planetary nebula **NGC 7662** or Caldwell 22, lies 26′ south-southwest of 13 Andromedae in the same low-power field. At first glance it masquerades as a 9th-magnitude star. My 4.1-inch refractor at 87× shows a small, round, faintly bluish disk with an elusive, star-like point in one side. Increasing the magnification to 153×, I detect a slightly oval shape elongated northeast to southwest. Occasionally there are hints of a small, slightly darker center.

Planetary nebulae were so named because many of them resemble the blue-green disks of the planets Uranus and Neptune. The color of NGC 7662 becomes more obvious with larger scopes, and it has earned the

STARQUEST 2007 Seen B4 @ Coyle

A beautiful spiral galaxy strongly tipped to our view, NGC 7331 happens to lie in nearly the same line of sight as three small and remote galaxies to its left (east). But each of them is no brighter than 14th magnitude and well beyond reach of a small telescope (visually).

10/12/2007

found @ Star Quest 2007 solid blue-green

Nicknamed the Blue Snowball by observers, the planetary nebula NGC 7662 has a flowerlike structure in images made with long-focus telescopes. This CCD view by William McLaughlin was taken with a 12½-inch Ritchey-Chrétien reflector.

name Blue Snowball. When it comes to planetary nebulae, there seem to be blue-eyed and green-eyed observers. Some consistently see certain planetaries as greenish, while others describe the same nebulae as some shade of pure blue. To me the Blue Snowball is a cross between sky blue and turquoise. What color do you see?

While you're in the area, try for the rather challenging galaxy **NGC 7640.** Given its visual magnitude of 11, you'd think this would be a fairly easy galaxy to spot. But the magnitude figures quoted for galaxies refer to their overall brightness, meaning that NGC 7640 would shine with the light of an 11th-magnitude star if its glow were gathered into a single point. Since the galaxy's light is actually spread over a fairly large area, it has a very low surface brightness.

To look for NGC 7640, scan 58′ south-southwest of the Blue Snowball and you'll encounter a 6th-magnitude orange star. Continue in the same direction for another 51′ to search for this elusive galaxy. It's easy to pinpoint because it sits enshrined in a triangle of 11th-magnitude stars. Ironically, the same stars that frame it so nicely are distracting enough to make this dim galaxy more difficult to discern.

With the 4.1-inch scope I find the galaxy easiest to pick out at 68×. It appears extremely dim, elongated north-northwest to south-southeast, and approximately 6′ long by one-fifth as wide. A 13th-magnitude star lies just off the south end. The galaxy also shows a barely brighter, elongated core. If you snare this darkling star city, your eyes will be collecting photons from a galaxy that has gone unnoticed by many observers with much larger scopes.

Next we'll visit the pretty triple star **Hough 197,** located 3.3° southwest of NGC 7640. If you were unable to track down that galaxy, try star-hopping from Omicron (o) Andromedae. Omicron is the star shown on our all-sky chart, right above the h in *Zenith.* You can spot it in the sky by noting that the right-hand side of the Great Square points northward to it. As our detailed map shows, Ho 197 lies 4.5° south-southeast of Omicron.

I can spot all three components of Ho 197 at 17× through my 4.1-inch scope, and they are widely split at 47×. The 7.9-magnitude, yellow-white primary has a 9.7-magnitude secondary 55″ to the west and a 10.2-magnitude tertiary 38″ to the northwest. They form a nice isosceles triangle, small and squat. While the companion stars seem to have slightly different hues, the colors are too subtle for me to define. From brightest to dimmest, the stars have spectral types of F5, F2, and F8, so the secondary should be a little whiter than the primary, and the dimmest star should be a little yellower. Can you see the difference?

From Ho 197 we'll sweep 3.6° west-southwest to locate the beautiful double star **h975** in the southeastern corner of the constellation Lacerta. The 5.7-magnitude, blue-white primary is the brightest star in the area, and its 9th-magnitude companion 53″ to the west-southwest has a nicely contrasting orange hue. The h in the name means this is one of John Herschel's discoveries early in the 19th century.

If looking for NGC 7640 gives you a headache, you'll appreciate the much easier galaxy **NGC 7331** (Caldwell 30) in Pegasus. To find it, place h975 at the northern edge of a low-power eyepiece field and move 4° westward. You should spot an

Beta (β) Pegasi, the 2.5-magnitude star marking the northwestern corner of the Great Square of Pegasus, is the jumping-off place for locating a planetary nebula, a pair of nice galaxies, and two multiple stars. The dim stars of this region occupy a small patch of sky that is practically overhead for midnorthern latitudes as soon as it gets dark on a November night.

attractive pair of 6th-magnitude orange stars. NGC 7331 lies 1.2° to their south.

NGC 7331 is the brightest galaxy in Pegasus. Although French astronomer Charles Messier missed this galaxy when compiling his famous catalog in the late 18th century, NGC 7331 is brighter than some of the galaxies on Messier's list. It has about the same structure and inclination toward us as the Andromeda Galaxy (see page 156). But since it is about 20 times farther away, we see it as a much smaller and dimmer version of its famous cousin.

Through my little refractor at 87×, NGC 7331 is fairly bright. It appears 5′ long and about 1′ wide, elongated north-south. The galaxy brightens toward the long axis and has a nearly stellar nucleus.

All our objects lie along a gentle arc that sweeps across three constellations while spanning only 13° of sky. You could almost hide the whole area with a fist held at arm's length, yet spend an whole evening enjoying the sights it covers.

Autumn Sights

Object	Type	Size/Sep.	Mag.	Dist. (l-y)	RA	Dec.
NGC 7662	Planetary	32″ × 28″	8.3	3,900	23^h 25.9^m	+42° 32′
NGC 7640	Galaxy	10.5′ × 2.0′	11.3	28 million	23^h 22.1^m	+40° 51′
Ho 197	Triple star	55″, 38″	7.9, 9.7, 10.2	500?	23^h 11.4^m	+38° 13′
h975	Double star	53″	5.7, 9.0	800?	22^h 55.7^m	+36° 21′
NGC 7331	Galaxy	10.5′ × 3.7′	9.5	49 million	22^h 37.1^m	+34° 25′

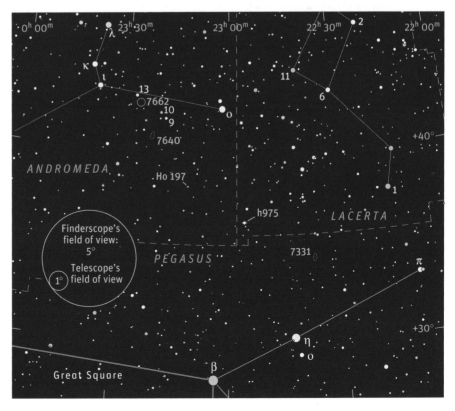

Hunting Down the Helix

WITHOUT A DOUBT, the Helix Nebula is one of the most beautiful objects in the heavens. It earns its name from a double-ringed appearance on photographs, like looking down at two coils of a spring. Long-exposure photos show nebulous fingers pointing inward from the ring toward the central star. The Hubble Space Telescope captured stunning close-ups of these radial filaments, showing each with a cometlike head and a gossamer tail. (Go to the HST Web site at http://hubblesite.org/gallery and search for Helix.)

Also known as NGC 7293 and Caldwell 63, the **Helix Nebula** is one of the nearest and brightest of the class of objects known as planetary nebulae. A planetary nebula is born of an aging star that exhausts its nuclear fuel and sheds its outer layers into space. As the nebula expands, the core of the star is exposed. This hot, dense cinder will evolve into a white-dwarf star and then slowly cool over billions of years.

Observationally, the Helix is one of the easiest of bright planetaries and one of the most elusive. It can

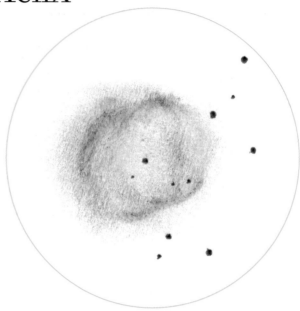

While observing at the Madison Astronomical Society's dark-sky site near Brooklyn, Wisconsin, Bill Ferris made this sketch of the Helix on September 27, 1995, with a 10-inch f/4.5 Newtonian reflector, Lumicon O III filter, and Meade super-wide-angle eyepiece at 63x. "The transparency was exceptional that night," he writes. North is up. Compare Bill's sketch to Gary L. Stevens' photograph at left.

For this striking view of the Helix, Gary L. Stevens took his Takahashi 152-millimeter refractor to Mount Pinos, California, on July 4, 2000. This image is a composite of seven white-light exposures at f/6 and three more through color filters using an SBIG ST-8E CCD camera. The total exposure time was 100 minutes. The field here is ½° across.

be seen in binoculars and yet remain invisible in large telescopes. These claims may seem contradictory, but they can be explained by the nebula's low surface brightness. If the light of the Helix were gathered into a single point, it would shine with the light of a 7.3-magnitude star. But this light is spread over a rather large area of the sky with about half the angular diameter of the Moon! Low powers will help concentrate the nebula's light, and wide fields will show plenty of surrounding dark sky for contrast. That's why the Helix is a good small-scope target.

With such an elusive quarry, you need to hunt the Helix carefully — being sure that your scope is aimed at the correct spot in the sky. The Helix is never very high in the sky from my semirural site in upstate New York, but I can sometimes see an adjacent star, Upsilon (υ) Aquarii, with the unaided eye. Spotting Upsilon will make your nebula search simple. But if your sky is too bright to show that star, try starting at 3.3-magnitude Delta (δ) Aquarii. From Delta, look 4° southwest for 4.7-magnitude 66 Aquarii. It should fit in the same finder field with Delta and shine with an orangish light through the telescope. Continue along that line another 2.8° and you will come to yellow-white Upsilon, the brightest star in the area at magnitude 5.2. The Helix sits 1.2° west of Upsilon and is best sought with a low-power eyepiece. Two 10th-magnitude stars lie halfway between, helping pinpoint the Helix in two easy jumps.

I can pick out the Helix with 50-millimeter binoculars from home. At first glance a small telescope shows only a featureless oval disk, but keep looking. The Helix Nebula yields up its details only to the patient observer.

On nights of good transparency, my 4.1-inch (105-mm) scope at low to medium power shows a 14′-by-11′ oval glow elongated northwest to southeast. The center is slightly darker than the rim. Averted vision (directing your gaze to one side; see page 17) can help you see the nebula better, and O III and narrowband light-pollution filters often work well. At high power, without a filter, some faint stars can be seen embedded in the nebulosity.

But what about all the beautiful features visible in photographs? While you can't observe the wealth of colorful detail that appears in deep images, there is more to see than you might think. I've read many observing reports and asked folks to tell me the smallest instrument they've been able to detect various features in. Let's see what they have to say.

Amazingly enough, Michael Bakich writes, "From our dark-sky site 50 miles east of El Paso, three of us saw the Helix with the naked eye on August 26, 2000." And in Western Australia, where the Helix passes almost overhead, Maurice Clark and his friends have used its naked-eye visibility as a guide to sky conditions. Others have found it through 6 × 30 finders and described it as easy with small binoculars. The annularity of the Helix has been noted in 2.4-inch telescopes. The central star has been spotted in a 6-inch.

After this we are getting beyond the realm of small telescopes, but I know from letters that many readers actually use fairly sizable apertures. Here are a few challenges for larger instruments.

Through 8-inch scopes, the Helix has been seen to shine with a blue-green glow. The human eye is more sensitive to this color than to the reddish hue that dominates photographs. This is why an

oxygen III filter improves the view so much; it passes the blue-green light given off by doubly ionized oxygen (designated O III) while blocking the colors common to many sources of light pollution. Irregularities in the brightness of the ring also start to show up with an 8-inch scope.

The two "coils of the spring" have been clearly seen by Bill Ferris of Flagstaff, Arizona, using a 10-inch reflector and O III filter. Like many planetaries, the Helix has a faint outer halo, but it doesn't show on most photographs. The brightest section emanates from the southeastern edge, spirals out clockwise, and fades in the north. While this halo has been suspected in a 10-inch, a 20-inch is the smallest scope I know of in which it has been described as apparent.

Another tough target is a 16th-magnitude galaxy embedded in the northwestern edge of the annulus. It is located 1.2′ south of the prominent 9.9-magnitude star at that end and has been seen in a 17½-inch scope. And what about those radial spokes? The streamers have been seen in a 16-inch and hints of the cometary globules at their ends with a 22-inch.

Many of these observations were made by very experienced deep-sky enthusiasts. Some of them observe from darker skies and more southerly latitudes than most of us enjoy, but their accomplishments give us something to strive for. When trying to nab any of these features, choose nights of good transparency and the darkest sky you can find.

The Helix Nebula

Designations	NGC 7293, C63
Right ascension	$22^h\ 29.6^m$
Declination	−20° 48′
Angular size *	16′ × 12′
Total visual magnitude	7.3
Mag. of central star	13.5
Distance in light-years	300

* The angular size includes some of the outlying halo.

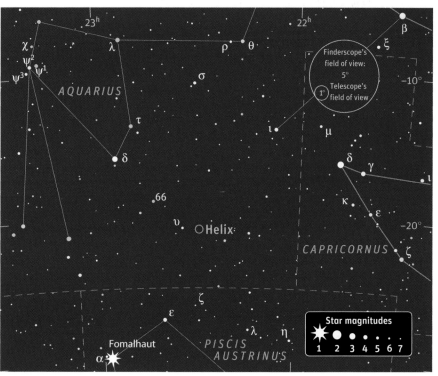

Treasures of the Royal Couple

HIGH IN THE AUTUMN SKY the Milky Way east of Cygnus passes through the constellations Cepheus, the King, and Cassiopeia, his Queen. They're nearly overhead when you face north around the time of our evening sky map on page 30.

Many deep-sky wonders for small telescopes reside in this royal realm. This time we'll look through the section of the Milky Way where Cepheus and Cassiopeia meet. You can use the map at right with your telescope to star-hop from the bright, landmark stars of Cassiopeia's **W**

Above: Compare the star patterns and clusters in this photograph by Akira Fujii with the map on the facing page. North is to the upper left. *Inset:* South of Mu Cephei sprawls the enormous, very dim nebula IC 1396. I was able to detect it using a 4.1-inch refractor at 17× in a not entirely dark sky where the faintest star visible to the naked eye was magnitude 5.9. This field is 3.6° wide.

pattern to each of our chosen sights.

Draw a line from bright Alpha (α) Cassiopeiae through Beta (β) and continue the line onward by a little more than the same distance. This will bring you to the open star cluster **M52**. It's just south of the 5th-magnitude star 4 Cassiopeiae. Both the cluster and this pretty orange star fit into the same low-power telescopic field of view. Binoculars or a large finderscope show M52 as a small, faint, nebulous patch of light. A 3-inch (76-millimeter) telescope begins to reveal some individual stars. (The bright star on the cluster's southwestern edge is not a true member of the group.) My 4.1-inch scope at 87× can resolve about 50 of the cluster's 200 stars and show the bright star's delicate yellowish orange tint. The cluster's center appears slightly hazy from the light of many more stars too dim to resolve individually.

A richer cluster lies 3° southwest of Beta Cassiopeiae, about halfway between the two 5th-magnitude stars Rho (ρ) and Sigma (σ). **NGC 7789** is visible through a good finderscope as a dim, round smudge. My 14 × 70 binoculars show many extremely faint pinpoints of light salted across a misty background. Through my 4.1-inch scope this cluster is a beautiful sight, rich in faint, 10th- through 12th-magnitude stars on a patchy background haze. An 8th-magnitude star highlights the western edge. NGC 7789 is quite old as open clusters go, with an age of about 1.7 billion years compared to M52's 60 million.

Draw a line from Beta Cas through NGC 7789 and continue on for twice that distance, and you come to an easily overlooked cluster in an area devoid of naked-eye stars. **Stock 12** is a large, sparse group about $1/3$° across. Using a 3.6-inch scope at 45× I can count about 45 or 50 stars. The brightest dozen are 8th and 9th magnitude. Five of the brighter ones in the northern part of the group form a curve with its concave side facing southwest. Following the curve southeast points the way to a nice, close pair of stars nearly matched in color and brightness. A smaller arc of three fairly bright stars is conspicuous south of the larger group; it curves in the same direction. The brightest member of Stock 12 is an 8th-magnitude luminary at the southwestern edge.

Moving farther west, look for the three naked-eye stars forming a small triangle at the southeastern corner of the King's constellation figure. The triangle's narrow point is the famous variable star **Delta (δ) Cephei,** prototype of the Cepheid variables — giants whose rates of pulsation tell their luminosities and therefore their distances. You can follow Delta Cephei's changes with the unaided eye. It brightens from magnitude 4.3 to 3.6 in about $1\frac{1}{2}$ days, then fades back to minimum

NGC 7789 (far left) and M52 (left) were photographed with identical equipment and exposures on the same night. In each case the field is 1° wide. *Using the map below:* The circles in the upper-left corner show the size of a typical finderscope's field of view (5°) and a typical telescope's view with its lowest-power, widest-field eyepiece (1°). This indicates how much of the map you'll see at a time. Next, figure out directions. North is up and east is left on the map. To find which way is north in your eyepiece view, nudge your telescope slightly toward Polaris; new sky enters the view from the north edge. Turn the map around to match. (If you're using a right-angle star diagonal at the eyepiece, it probably gives a mirror image. Take it out to see a correct image that will match the map.)

light in just under 4 days. You can compare Delta's brightness to that of Zeta (ζ), magnitude 3.4, and Epsilon (ε), magnitude 4.2, whenever you look up on a clear autumn night.

For the telescope, Delta Cephei is a beautiful double star that appears widely split at 20×. In my 4.1-inch refractor the primary (the variable) is yellow, and the 6.3-magnitude secondary star appears blue. They are 41 arcseconds apart.

Moving 6° west we come to Herschel's Garnet Star, **Mu (μ) Cephei.** It's uncommonly deep orange-red; a small telescope shows its color nicely. Mu Cep is one of the largest star visible to the naked eye — a red supergiant roughly 2.4 billion miles (3.8 billion kilometers) across. If it replaced our Sun, it would extend well beyond the orbit of Jupiter.

Mu Cep is a slow semiregular variable. It is easily visible to the unaided eye in suburban skies when at its maximum brightness of magnitude 3.4, but it may be a challenge at its minimum of 5.1.

Edge your telescope 1.3° south-southwestward from Mu Cephei to find two attractive multiple-star systems, **Σ2816** and **Σ2819** (Struve 2816 and 2819). Both fit into the same field of view at 50×, which is enough magnification to split all their visible components.

Struve 2819 is the dimmer system, but with magnitudes of 7.4 and 8.7 and a separation of 12″, the pair is easy to resolve in a 2.4-inch. The secondary star lies northeast of the primary.

Struve 2816 is triple. Two 8th-magnitude companions flank the 5.7-magnitude primary, one to the east-southeast and one to the north-northwest, with separations of 12″ and 20″, respectively.

All around them are a sparse star cluster and huge, dim emission nebula collectively known as IC 1396. The nebula may be dimly glimpsed in large binoculars or a low-power, rich-field telescope on a very dark, moonless night.

Now turn the page to explore more of Cepheus, full of treasures truly fit for a King.

Treasures of the King and Queen

Object	Type	Mag.	Dist. (l-y)	RA	Dec.
M52	Open cluster	6.9	4,600	23ʰ 24.2ᵐ	+61° 35′
NGC 7789	Open cluster	6.7	7,600	23ʰ 57.0ᵐ	+56° 44′
Stock 12	Open cluster	7?	—	23ʰ 35.6ᵐ	+52° 42′
δ Cephei	Variable, double star	3.6–4.3, 6.3	1,000	22ʰ 29.2ᵐ	+58° 25′
μ Cephei	Variable red star	3.4–5.1	3,000?	21ʰ 43.5ᵐ	+58° 47′
Σ2819	Double star	7.4, 8.7	—	21ʰ 40.4ᵐ	+57° 35′
Σ2816	Triple star	5.7, 8.0, 8.1	1,200	21ʰ 39.0ᵐ	+57° 29′

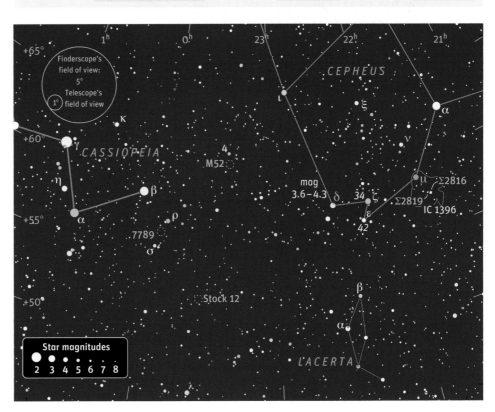

A Regal Treasure-Trove

ON FALL EVENINGS, the constellation Cepheus, the King, lies temptingly high overhead. Its southern reaches are bathed in the riches of the Milky Way, where deep-sky wonders abound.

Let's begin in the southwestern corner of Cepheus with one of the sky's most captivating sights. Here we see a galaxy and an open star cluster just ²/₃° apart — close enough to be embraced by a single low-power view. Their nearness is an illusion, of course. The galaxy is nearly 5,000 times more distant than the cluster, giving us a sense of the profound depths of space.

The members of this celestial odd couple — the star cluster NGC 6939 (top right) and the galaxy NGC 6946 — can be viewed simultaneously in a low-power telescope. But Jerry Lodriguss's photograph brings out intricacies no human eye can see. He made the 70-minute exposure on gas-hypered Fujicolor Super G800 film at Massai Point, Arizona. It was taken on May 27, 1995, with an Astro-Physics 130-millimeter refractor at f/8.

Open cluster **NGC 6939** is the brighter of the two. To pinpoint it, place Theta (θ) Cephei in a low-power field and drop 2.3° south. Once you spot the cluster, seek out **NGC 6946** (Caldwell 12) to the southeast.

With my 4.1-inch refractor at 47× they make a beautiful pair. NGC 6939 shows many very faint pinpoints over a hazy background. NGC 6946 is a bit larger, slightly oval, and appears as a nearly uniform hazy glow with only a slight increase in brightness toward the center. A small, squat triangle of 10th- through 12th-magnitude foreground stars is partly embedded in the southern part of the galaxy.

When I boost the magnification to 87× with a wide-angle eyepiece, the duo just squeezes into the same field. The cluster is very rich in faint stars and appears 8′ across. Several exceedingly faint stars can be seen scattered across the galaxy, which looks about 10′ by 8′ and elongated northeast to southwest. Under dark skies with his 4-inch scope and much patience, noted observer and *Sky & Telescope* contributing editor Stephen James O'Meara has spotted traces of spiral structure in this galaxy.

Eight supernovae have been observed in NGC 6946, more than in any other galaxy. The first appeared in 1917 and the last in 2004, within a human lifespan. Such a prolific galaxy is worth keeping an eye on! Its neighbor is also unusual. Most open clusters are loosely bound and lose hold of their stars within a few hundred million years, but NGC 6939 is relatively ancient at 2 billion years old.

Now we'll march eastward across Cepheus through a succession of small open clusters. The first is **NGC 7160,** located about two-fifths of the way from Nu (ν) to Xi (ξ) Cephei. The cluster is easy to identify because its two brightest stars, 7th and 8th magnitude, form the wide pair known as South 800. In 14 × 70 binoculars, the double and a few fainter stars look like a stubby caterpillar with glowing eyes. My little refractor at 87× shows six fairly bright stars in a distinctive shape. South 800 and a star to the northeast make an arrowhead, while a little curve of stars to the west-southwest forms a bent shaft. A dozen fainter stars are strewn about the area.

Our next cluster is **NGC 7235**, situated 25′ northwest of Epsilon (ε) Cephei. It is visible in 14 × 70 binoculars as a single star offset in a small, dim patch of light. My refractor at 87× shows five moderately bright to faint stars plus seven very faint stars in a little clump. The three brightest mark the corners of a skinny east-west triangle that almost spans the group.

Now center a low-power eyepiece on the striking yellow and blue double Delta (δ) Cephei (see page 146). Move 2.4° east to a blue-white star of 6th magnitude.

Notice that the star is parked at the pointy end of a slender triangle. One corner of its short base is marked by a star of similar brightness, the other by the double star OΣ 480. **NGC 7380** sits just east of this pair. At 87×, I see a pretty group of 30 moderately bright to faint stars in a shape like a witch's hat. The hat's large floppy brim forms the eastern side, and OΣ 480 marks its pointed top. The cluster has a hazy background, but am I seeing unresolved stars or nebulosity? NGC 7380 is, in fact, embedded in a very faint nebula, Sharpless 2-142. At 28×, a green oxygen III filter seems to slightly enhance and extend the haziness compared to an unfiltered view. If you have a light-pollution filter, can you detect the nebula with a small scope?

Our final stop will be in far eastern Cepheus, where several interesting targets lie close together. The brightest is **NGC 7510.** To find it, look for the naked-eye star 1 Cassiopeiae. A low-power view will show a 6th-magnitude star to its east. Centering that and sweeping 1.2° north will take you to a conspicuous knot of stars. My little scope at 153× shows a small group of a dozen faint to very faint stars. The brightest of these form a bar running east-northeast to west-southwest; the rest lie north of the bar and turn the cluster into a triangle.

While we're visiting Cepheus, the King, it's only fitting that we look in on one of the clusters reported by Ivan R. King in the Harvard College Observatory Bulletin a half century ago. Four lie in the constellation Cepheus, none very striking in a small scope. **King 19,** however, has the redeeming quality of being easy to locate. It is a faint knot of stars 1/3° west of NGC 7510 in the same low-power field. At 153×, I count 10 faint to very faint stars gathered in a loose, irregular group. The brightest make a triangle in the cluster's eastern side, a three-star line in the western side, and a solitary point between the trios.

Markarian 50 is a smaller but brighter knot 1/2° east and a little south of NGC 7510. At 127×, I see five stars in a tiny curve reminiscent of a miniature Corona Borealis. Avid deep-sky enthusiast and author Tom Lorenzin calls this group the Tiny Tiara. Mrk 50 sits off the northwestern edge of the brightest arc of the large nebula **Sharpless 2-157.** Although barely detectable without a filter, the nebulosity is fairly obvious with an O III filter at 17×. It is curved concave westward, wider in the south, and about 1° long north-south. Lorenzin nicknamed this the Californietto Nebula for its resemblance to the much larger California Nebula in Perseus (see page 35).

We've spanned Cepheus from west to east, sampling the kingly treasures that bejewel this swath of sky. These riches belong to any observer with a clear, dark night and the will to behold them. But we haven't finished with this region of the sky. Turn the page to discover eight delightful open star clusters in Cassiopeia, the Queen.

Some of Cepheus' Speckled Wealth

Object	Type	Mag.	Size	Dist. (l-y)	RA	Dec.
NGC 6939	Open cluster	7.8	7′	3,900	20ʰ 31.5ᵐ	+60° 39′
NGC 6946	Galaxy	8.8	11′ × 10′	19 million	20ʰ 34.9ᵐ	+60° 09′
NGC 7160	Open cluster	6.1	7′	2,600	21ʰ 53.8ᵐ	+62° 36′
NGC 7235	Open cluster	7.7	4′	9,200	22ʰ 12.5ᵐ	+57° 17′
NGC 7380	Open cluster	7.2	12′	7,200	22ʰ 47.4ᵐ	+58° 08′
NGC 7510	Open cluster	7.9	4′	6,800	23ʰ 11.1ᵐ	+60° 34′
King 19	Open cluster	9.2	6′	6,400	23ʰ 08.3ᵐ	+60° 31′
Mrk 50	Open cluster	8.5	2′	6,900	23ʰ 15.2ᵐ	+60° 27′
Sh2-157	Diffuse nebula	—	60′ × 10′	6,900	23ʰ 16ᵐ	+60.3°

Angular sizes are from catalogs or photographs; most objects appear somewhat smaller when a telescope is used visually.

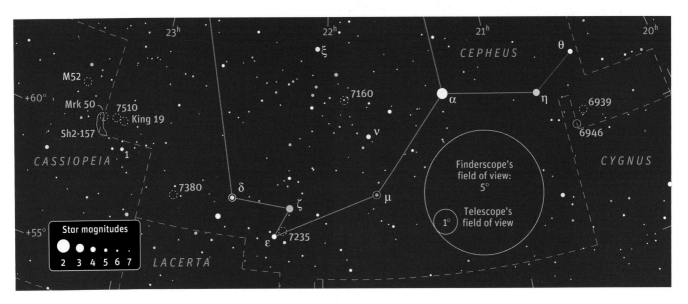

Cassiopeia's Chair

THE CONSTELLATION CASSIOPEIA, the Queen, sports five bright stars arranged in a memorable pattern. Facing north on a June evening, we'd see it as a **W**-shaped asterism brushing the horizon. But now, a half year later, Cassiopeia rides high in the north and is flipped upside down like an **M** nestled in the misty band of the Milky Way.

While it's difficult to envision the Queen herself among these stars, some skygazers like to add the star Kappa (κ) Cassiopeiae to create her chair, or throne (see the pair of sketches at lower right opposite). Kappa is a prime example of the old saw "You can't judge a book by its cover." Although it looks dimmer than the other throne stars, this intensely hot, blue supergiant is a whopping 480,000 times more luminous than our Sun. The glory of the star is dimmed by interstellar dust and by the fact that it lies some 3,300 light-years away. It doesn't help that it emits much of its energy in the form of unseen, ultraviolet light.

Numerous star clusters lie within 5° of this remarkable star. Since none belong to the well-known Messier catalog, they are seldom visited — yet many are within the grasp of a small scope.

The hardest part about observing these clusters is picking them out against the starry backdrop of the Milky Way. This is certainly true for our first three star clusters. They are centered ⅓° north-northwest of Kappa.

King 14 lies closest to Kappa and is the largest of the trio. Through my 4.1-inch (105-millimeter) refractor at 87×, it appears 8′ across with 20 faint to very faint

stars. Ivan R. King announced this cluster in 1949, in Harvard Observatory *Bulletin* 919. **NGC 146** is just northeast of King 14, looking a little smaller with about 15 stars. The two clusters are splayed out from each other like a pair of narrow moth wings. **NGC 133** seems the smallest of the three, lying north-northwest of King 14. It consists of five 10th- and 11th-magnitude stars in a **Y** shape, along with a few even fainter stars.

These clusters aren't conspicuous, but together they form an interesting pattern. If King 14 and NGC 146 are spread wings, NGC 133 could be a tail. On the opposite side of the wings is a wedge-shaped group of stars that is not part of these clusters. I see it as the profile of a bird's head capped by a 9th-magnitude star, reminding me of the crest of a pileated woodpecker.

Now that you have a feel for what to expect when hunting clusters in a crowded setting, let's roam a bit farther afield. First we'll try **NGC 225,** which lies roughly halfway from Kappa to Gamma (γ) Cassiopeiae. My 4.1-inch scope at 59× reveals 20 moderately bright to fairly faint stars in an irregular group 12′ in diameter. These stand out fairly well in a relatively sparse area of the Milky Way. NGC 225 is one of several "remarkable appearances" recorded by Caroline Herschel, who discovered it in 1784 with her 4.2-inch Newtonian comet seeker in southern England.

Next we'll move to **NGC 129,** a fairly large open cluster 2.7° south of Kappa. To pin down its location, notice that the cluster is situated along a line connecting Gamma to Beta (β) Cassiopeiae. At 68× I see three moderately bright stars in a right triangle and one more to the north-northeast, outlining a skinny kite. Some 30 faint stars join the kite in a loosely scattered bunch about 20′ across. The star at the northwestern point of the triangle is the Cepheid variable **DL Cassiopeiae.** This pulsating yellow giant ranges between magnitudes 8.6 and 9.3, the rise to maximum being much swifter than the descent to minimum. A full cycle takes eight days.

NGC 7790 and **NGC 7788** make a pretty double cluster through a small telescope. They sit 4.5° west-southwest of Kappa and are a shorter star-hop from Beta. If you put Beta at the south-southeastern edge of your finderscope's field, the clusters should be visible in a low-power eyepiece on your main scope. For those using a unit-power finder (also called a reflex finder), note that NGC 7790 is at the right angle of a triangle that has Kap-

Cassiopeia's bright stars punctuate a fairly rich section of the northern Milky Way. Celestial north is up; turn the book upside down to match the star's orientation high in the northern sky.

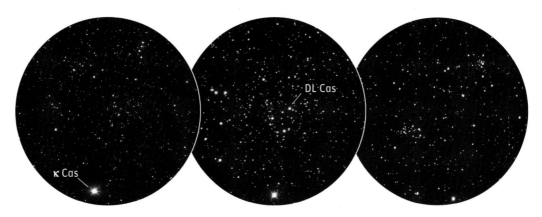

George Viscome captured six of Cassiopeia's clusters in close-ups with his 14½-inch f/6 reflector. *Left:* King 14, NGC 146, and NGC 133. *Middle:* NGC 129 with DL Cassiopeiae. *Right:* NGC 7790 (lower left) and NGC 7788 (upper right). North is up.

pa and Beta at the other two corners.

At 87×, I see NGC 7790 as a teeming group of faint to extremely faint stars. Its concentrated core is elongated east to west. The three brightest stars arc along the edge from north to west. An extended, misty halo about 10′ across fades outward from the core. NGC 7788 shares the field of view. One 9th-magnitude star has a host of very faint companions crowded tightly nearby. Some 9th- to 10th-magnitude stars reach out from the group, making a **V** that opens to the southeast.

Our final cluster will be **NGC 381,** which is easiest to get to from Gamma Cassiopeiae using a low-power eyepiece. Putting blue-white Gamma near the southern edge of the field of view, you should see a yellow 6.4-magnitude star to the north. Now place the yellow star at the western edge of your field and look for a yellow-white 5.8-magnitude star to the east. Moving the yellow-white star to the western edge should position NGC 381 near the center of your view. (To keep your directions straight, remember that stars drift from east to west through the field of an undriven telescope.)

In my little refractor at 87×, NGC 381 is a small, misty cluster enshrined in a pentagon of faint stars. Several specks of light glimmer faintly across the haze, and increasing the magnification to 127× teases out other dim suns. A wavy line of 11th- and

12th-magnitude stars trails northward from the cluster. In the eyes of California amateur Joe Bergeron, it looks like a balloon on a string.

I may have run out of room in this essay but certainly not out of clusters. A total of 39 open clusters lie within 5° of Kappa Cassiopeiae, about half of these being visible through a small telescope. Don't be surprised if you stumble across a few while searching for the little-known gems listed below.

Open Star Clusters in Cassiopeia

Object	Type	Mag.	Size	Dist. (l-y)	RA	Dec.
King 14	Open cluster	8.5	7′	8,500	0ʰ 31.9ᵐ	+63° 10′
NGC 146	Open cluster	9.1	6′	9,900	0ʰ 33.1ᵐ	+63° 18′
NGC 133	Open cluster	9.4	7′	2,100	0ʰ 31.3ᵐ	+63° 21′
NGC 225	Open cluster	7.0	12′	1,700	0ʰ 43.5ᵐ	+61° 47′
NGC 129	Open cluster	6.5	21′	6,600	0ʰ 30.0ᵐ	+60° 13′
DL Cas	Variable star	8.6–9.3	—	6,600	0ʰ 30.0ᵐ	+60° 13′
NGC 7790	Open cluster	8.5	17′	10,800	23ʰ 58.4ᵐ	+61° 12′
NGC 7788	Open cluster	9.4	9′	7,700	23ʰ 56.7ᵐ	+61° 24′
NGC 381	Open cluster	9.3	6′	3,100	1ʰ 08.3ᵐ	+61° 35′

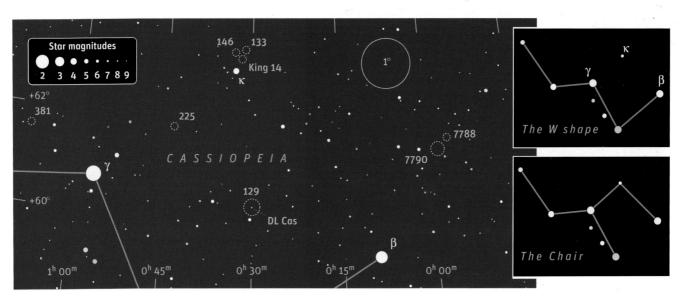

File Under *W*

ON THIS MONTH'S all-sky star chart on page 31, the prominent **W** traced by Cassiopeia's brightest stars sits slightly above center. The bright and easily recognized pattern will serve as our guidepost to some of the deep-sky delights lying beneath, or rather south of, this famous asterism.

The beautiful double star **Eta (η) Cassiopeiae** will be our first stop. It is visible to the unaided eye one-third of the way from Alpha (α) to Gamma (γ) Cassiopeiae and south of a line connecting them. Eta's components weigh in at magnitudes 3.5 and 7.4 and have been variously described by well-known double-star observers of the 19th century. English clergyman Thomas W. Webb called the pair yellow and pale garnet, Wilhelm Struve saw them as yellow and purple, and John Herschel and James South recorded them as red and green! With my 4.1-inch (105-millimeter) refractor, I see the brighter star as yellow and the dimmer one as red-

Two of this month's colorful little clusters, NGC 436 *(above)* and NGC 457 *(facing page)*, were recorded by George R. Viscome with his 14-inch Newtonian reflector at Lake Placid, New York. Each view is ²/₃° wide. I see NGC 457 as the Dragonfly (note the "eyes").

dish orange, in good agreement with their spectral types of G0V and M0V. The pair is split at 29×, and widely so at 36×.

The suffix "V" in each star's spectral type indicates it, like our Sun, is fuzing hydrogen in its core. The secondary is one of the very few observable red dwarfs that are paired with a naked-eye star. Shining at about 6 percent of our Sun's luminosity, this ruddy star is visible through small scopes only because it is relatively nearby at 19 light-years. If you're wondering how our Sun would appear at that distance, simply cast your gaze upon the primary — a very Sun-like star.

Just 1.3° south-southeast of Eta, we find the nebula **NGC 281** and its associated star cluster, IC 1590. The nebula makes a nice isosceles triangle with Eta and Alpha (α) Cassiopeiae. Its faint, irregular glow is easy to spot with my refractor at 68× and occupies about ¹/₃°. A narrowband light-pollution filter improves the view a bit, while a green oxygen III filter helps a little more.

The nebula was discovered in the late 19th century by American astronomer Edward Emerson Barnard. Only later did Guillaume Bigourdan of France spot the open cluster within it — and little wonder! IC 1590 is not at all obvious through the scope. It is redeemed, however, by Burnham 1, the pretty multiple star at its core. Burnham 1 is the brightest star enshrined in the nebula. At 87× I see three close components, all bluish white, with magnitudes of 8.6, 8.9, and 9.7. If the air is very steady, try examining the brightest of this trio at a magnification of 200× or more. It has a 9.3-magnitude attendant a scant 1.4″ to the east — a severe test for small telescopes. This fourth component turns Burnham 1 into a tiny trapezium ensconced in its own miniature Orion Nebula. Like its larger and more famous counterpart, the trapezium of Burnham 1 supplies much of the radiation that compels its parent nebula to shine.

A line from Epsilon (ε) through Delta (δ) Cassiopeiae points toward Phi (φ) Cas, which is visible to the naked eye under moderately dark skies. Phi is a lovely double star that can be split even with binoculars. Its components appear to be the two brightest stars of the open

The soft glow of NGC 281 can't be perceived as truly colorful by human eyes, but its vivid red hue dominates this photographic image. Sean Walker used an 8-inch telescope at f/6.3 for this 55-minute exposure on Kodak E200 film. North is up and the field 1° wide.

cluster **NGC 457,** though they may actually be foreground objects. In my little scope at 87×, the 5th-magnitude primary appears yellow and the 7th-magnitude secondary looks blue-white. About 45 fainter stars are seen in a cluster that inspires imaginative flights of fancy.

In *1000+: The Amateur Astronomer's Field Guide to Deep Sky Observing,* author Tom Lorenzin calls it the ET Cluster after the cute extraterrestrial in the movie E. T. Lorenzin writes, "ET waves his arms at you and winks!" The two bright stars form ET's eyes, sprinklings of stars northeast and southwest are his outstretched arms, and his feet extend northwest.

Those big eyes lead quite naturally to the nickname of the Owl Cluster, which was created by David J. Eicher, now editor of *Astronomy* magazine. In a similar vein, I always picture NGC 457 as the Dragonfly. California amateur Robert Leyland is somewhat more creative. "Two wings of stars spread from either side of the cluster," he says, and they "give it the look of an F/A-18 jet on afterburners, with Phi Cas being the engines."

Its alternate designation is Caldwell 13. In his book *Deep-Sky Companions: The Caldwell Objects,* Stephen James O'Meara gives us this haunting impression of the view in his 4-inch scope: "The cluster's bright 'eyes' seem to pierce the night with the fiery gaze of a specter emerging from the dusty cobwebbed corridors of space. The ghost's clothes hang in tatters from skeletal limbs."

Behind my Dragonfly, 1/2° northwest of his tail, the open cluster **NGC 436** is a little ball of fluff at low power. At 28× in my 4.1-inch refractor I can make out only a few faint pinpricks of light in the fuzz. Upping the power to 153× teases out a dozen faint to very faint stars embedded in haze. Most of the brighter stars are arranged in diverging arms about 4' long. Leyland speculates that NGC 436 might be the target of his jet fighter, which is aimed in its direction.

Now draw an imaginary line from Delta through Chi (χ) Cassiopeiae; continue for almost 2½ times that distance again to reach the pretty open cluster **Stock 4.** At 68× I see a group of about 50 faint stars in a more or less rectangular bunch, about 20' by 12', that runs west-northwest to east-southeast. A bar of fairly bright to faint stars lies just northeast of the cluster and runs northwest to southeast. A widely spaced pair of 8th-magnitude stars can be seen to the east.

Stock 4 is one of the clusters discovered by Jurgen Stock and initially published in the first edition of the *Catalogue of Star Clusters and Associations* (Prague, 1958). So far it has received little attention from either amateur or professional astronomers. Since today's large printed and software atlases often include some of the sky's more obscure objects, many pleasant discoveries such as this await the notice of curious observers.

Now drop 1.8° south-southeast to another attractive group, **NGC 744.** At 87× my little scope shows 15 faint to very faint stars over haze. Disconnected star bunches lie southwest, south, east-southeast, and east-northeast.

Attractions on High

Object	Type	Mag.	Size/Sep.	Dist. (l-y)	RA	Dec.
η Cas	Double star	3.5, 7.4	13"	19	0ʰ 49.1ᵐ	+57° 49'
NGC 281	Diffuse nebula	8.0	28' × 21'	9,600	0ʰ 53.0ᵐ	+56° 38'
NGC 457	Open cluster	6.4	13'	8.000	1ʰ 19.6ᵐ	+58° 20'
NGC 436	Open cluster	8.8	5'	9,800	1ʰ 16.0ᵐ	+58° 49'
Stock 4	Open cluster	—	20'	—	1ʰ 52.7ᵐ	+57° 04'
NGC 744	Open cluster	7.9	11'	3,900	1ʰ 58.5ᵐ	+55° 29'

Angular sizes are from catalogs or photographs; most objects appear somewhat smaller when a telescope is used visually.

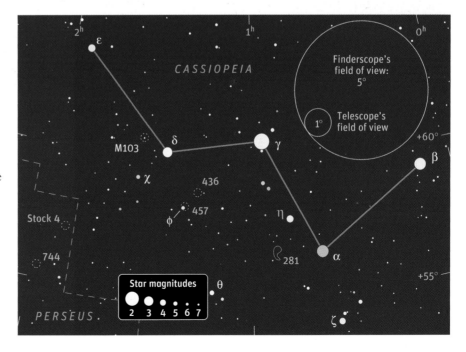

A Queen's Ransom

CASSIOPEIA, THE QUEEN, rides high across the northern sky in the evening at this time of year. Her W-shaped star figure (currently upside down) ranks with Orion and the Big Dipper as one of the best-known asterisms in the sky. Look for the **W** high in the north on the all-sky map on page 31.

The Queen's mythological figure sits on a starry throne with her foot marked by Epsilon (ε) Cassiopeiae, her neck by Alpha (α), and her slightly bowed head by dimmer Zeta (ζ). In this essay we'll use the familiar stars of Cassiopeia to help us find a wealth of royal jewels.

Along one line of the **W** lies an additional fairly bright star, **Eta (η) Cassiopeiae.** Look for it about a third of the way from Alpha Cas to the **W**'s middle star, Gamma (γ). Shining at magnitude 3.4, Eta is an easy find in suburban skies. It's a lovely double star that can be split at a magnification of 30× and resolved widely at 50×.

The bright component is a very Sun-like star with a spectral class of G0. Sure enough, it looks yellow-white through my 4.1-inch refractor. The secondary star, 13 arcseconds to the northwest, is a type-M0 red dwarf only ¹/₄₀ as bright at magnitude 7.4. To me it looks orange. Our Queen is wearing a pendant of topaz and hyacinth.

Above: M103 contains the bright triple star Σ131 on its northern edge. *Right:* NGC 654 is a compact little cluster with a 7th-magnitude yellow star just to its south-southeast.

The Eta Cas system is 19.4 light-years away, rather near for a naked-eye star. Its two suns revolve around each other with a period of about 480 years. They range from 35 to 110 astronomical units apart during each orbit; by comparison, Pluto's distance from the Sun averages 40 a.u.

The open star cluster **NGC 457** is also simple to locate. Draw a line from Epsilon through Delta (δ) Cas and continue on a bit less than half as far. That puts you close to 5th-magnitude Phi (φ) Cas, which shines right on the southeastern edge of NGC 457.

I particularly enjoy showing this cluster at public star parties. Open clusters are especially engaging when they conjure up some familiar shape. The arrangement of stars in this group has earned it various nicknames, including the Owl Cluster and the ET Cluster. For me, NGC 457 will always be the Dragonfly (see the previous page). Anyone can play this dot-to-dot game. What shape do you make of NGC 457?

There's more to this cluster (also known as Caldwell 13) than meets the eye. It may contain several thousand stars, but most are dimmed to invisibility by the group's remarkable distance of 8,000 light-years. If Phi Cas is really a member of this cluster it must be one of the intrinsically brightest stars known — more than 50,000 times as luminous as our Sun!

The only Messier object in this month's sky tour lies just 1° east-northeast of Delta Cassiopeiae. **M103** is a small open cluster that appears very triangular. A 3-inch scope shows a dozen stars, and a 4-inch reveals around 20. The three brightest form the triangle's northeastern side; the central one of these appears orange. The northern one is a wide triple, Σ131 (Struve 131), consisting of a 7.3-magnitude primary with faint companions 14″ and 28″ to the southeast, magnitudes 9.9 and 11.6, respectively.

Although Messier missed the open cluster **NGC 663**, it is a more enchanting sight in a small telescope than M103. Through my 4.1-inch at 87× it appears more

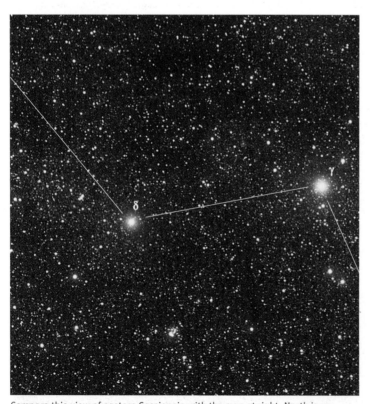

Compare this view of eastern Cassiopeia with the map at right. North is up.

order of decreasing brightness the three stars have spectral classes of *A3*, *F5*, and *G4*, suggesting colors of white, off-white, and pale yellow. Despite this I can't help but see the dimmest star as bluish. Color-contrast perception for double stars varies with their separations and brightnesses, as well as quirks of the observer and instrument. What do you see?

Sights Fit for a Queen

Object	Type	Mag.	Dist. (l-y)	RA	Dec.
η Cas	Double star	3.4, 7.4	19	0ʰ 49.1ᵐ	+57° 49′
NGC 457	Open cluster	6.4	8,000	1ʰ 19.1ᵐ	+58° 20′
M103	Open cluster	7.4	8,000	1ʰ 33.2ᵐ	+60° 42′
NGC 663	Open cluster	7.1	8,000	1ʰ 46.0ᵐ	+61° 15′
NGC 654	Open cluster	6.5	8,000	1ʰ 44.1ᵐ	+61° 53′
NGC 659	Open cluster	7.9	6,800	1ʰ 44.2ᵐ	+60° 42′
ι Cas	Triple star	4.5, 7, 8	140	2ʰ 29.1ᵐ	+67° 24′

than twice as wide, and I can count 40 stars ranging from fairly bright to extremely faint. The four brightest are 9th magnitude and occur in two widely spaced pairs. This attractive group, also called Caldwell 10, has a central concentration of very faint stars.

Both M103 and NGC 663 are part of a vast association of young stars known as Cassiopeia OB8. When you look at these stars you are gazing out to the Perseus Arm of our galaxy, the next spiral arm out from our Sun's (the Orion Arm).

The two smaller open clusters **NGC 654** and **NGC 659** lie less than 1° north-northwest and south-southwest, respectively, from NGC 663. Although faint, they are visible in a 2.4-inch scope as little nebulous spots. My 4.1-inch at 87× can resolve some of their stars.

NGC 654 is the prettier of the two, with a dozen faint stars sprinkled over a patchy, hazy background. A brighter, 7th-magnitude yellow star lies just outside the cluster's southern edge.

NGC 659 is a little smaller and dimmer. Seven faint stars in the rough shape of a fat pie wedge overlie a dim, hazy background.

All of our star clusters lie along the Queen's lap and legs. Her skirt is apparently adorned with patterns of tiny diamonds offset by an occasional colored gem.

Draw a line from Delta through Epsilon and continue for the same distance, and you come to **Iota (ι) Cassiopeiae** one of the nicest triple stars for amateur telescopes. Don't be confused by the two 9th-magnitude stars well to its east. Iota Cas has a 4.5-magnitude primary, a close and difficult 7th-magnitude secondary star currently just 2.6″ southwest, and an 8th-magnitude star 7″ east-southeast. I can separate all three at 87× through my 4.1-inch but find the view more pleasing at 153×. You'll need a night of good atmospheric seeing to resolve the close pair.

This system is 140 light-years away. In

How to use the map below: Note the circles in the upper-right corner. They show the size of a typical finderscope's field of view (5°) and a typical telescope's view with a low-power eyepiece (1°). This indicates how much of the map you'll see at a time. Next, figure out directions. North is up and east is left on the map. To find which way is north in your eyepiece view, nudge your telescope slightly toward Polaris; new sky enters the view from the north edge. Turn the map around to match. (If you're using a right-angle star diagonal at the eyepiece, it probably gives a mirror image. Take out the diagonal to see a correct image that will match the map.)

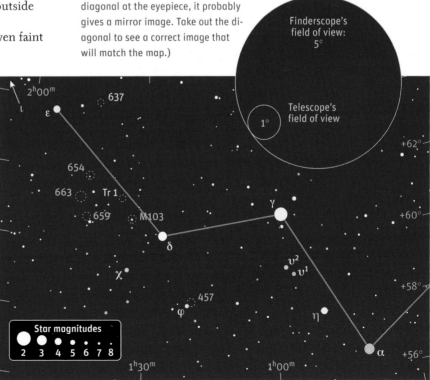

Local Calls

ON THIS MONTH'S all-sky star chart on page 31, Andromeda glitters prominently at the *zenith* (the center of the chart). Tucked away in this constellation is the brightest external galaxy visible from midnorthern latitudes — the **Great Andromeda Galaxy (M31).** It's a member of the Local Group, a loosely bound family of about 40 galaxies including our own. Although M31 is generally listed as the largest of the Local Group, two recent studies indicate that it may actually be a bit less massive than the one we call home.

The Andromeda Galaxy is just barely visible to the unaided eye as a small, elongated, nebulous haze even in suburban skies. When you gaze at M31, you are seeing light that started its journey around 2.5 million years ago. We find the earliest mention of it in Persian astronomer al-Ṣūfī's *Book of the Fixed Stars* (AD 964), where he referred to it as a "little cloud." To find M31, draw a line from 2.1-magnitude Beta (β) Andromedae through 3.9-magnitude Mu (μ) Andromedae and continue for that distance again.

M31 is a treat through any telescope. At first glance you may see only the galaxy's bright, oval heart, but careful study reveals much more. The faint, outer regions of M31 extend its length out to about 3°. They are easier to notice if you sweep your telescope slowly

The Andromeda Galaxy (M31) with its two elliptical companions M110 (also known as NGC 205) to the northwest and M32 to the south. Andromeda is the brightest galaxy in the northern sky and can be seen with the naked eye under dark-sky conditions.

across them, since the eye sees faint objects in motion better than stationary ones. These extensions are best seen through small, short-focus telescopes at 15 to 20×.

Examining the Andromeda Galaxy at higher powers unveils further details. At 47× in my 4.1-inch (105-millimeter) refractor, a stellar nucleus appears embedded in a small, bright, roundish inner core. This, in turn, is found within a larger, slightly dimmer, oval outer core about 25′ long.

M31 is a spiral galaxy inclined only 12.5° from an edge-on view. Looking closely, you will notice that the southeastern side of the galaxy fades slowly into the background sky, while the northwestern boundary ends quite abruptly. This sudden drop in brightness marks the position of a dark lane in M31's spiral arms. If the sky is very dark and transparent, you may spy a hint of faint fuzz beyond this dark lane — the next spiral arm outward.

There are at least a dozen satellite galaxies swarming near M31, two of which can be seen in the same low-power, wide-angle field of view. One is **M32,** located 24′ south of the Andromeda Galaxy's nucleus. M32 looks tiny in comparison to its enormous neighbor and might be overlooked in a casual view. Through my 4.1-inch scope at 68×, M32 is fairly bright, slightly oval, and fades outward from a conspicuous stellar nucleus.

The second companion is **M110,** located 37′ northwest of M31's nucleus. Although it is much larger than M32, it has a lower surface brightness and is a more difficult find. Through my little refractor, this galaxy appears about 20′ by 10′ elongated nearly north-south. It is fairly faint with a weakly brighter, oval core offset toward its northern end. Charles Messier discovered this galaxy in 1773 yet did not include it in his famous catalog of deep-sky objects. In recognition of Messier's discovery, NGC 205 has become popularly known as M110.

The third-largest member of the Local Group lies in nearby Triangulum. **M33,** the Pinwheel Galaxy (a nickname it shares with M101 and M99), is about half the diameter of our galaxy and 2.7 million light-years distant. It is a notoriously difficult target for the novice stargazer. This spiral galaxy lies almost face on so that it appears nearly the size of two full Moons laid side by side. With its light spread out over such a large area, M33 has a very low surface brightness and is easily washed out by moonlight or light pollution. Despite this, it can be seen through binoculars and has even been spotted with the unaided eye under very dark skies.

Look for the Pinwheel 4.3° west-northwest of 3.4-magnitude Alpha (α) Trianguli. To maximize your chances of seeing it, use a low-power eyepiece — one that gives you at least a 1° field of view. Try sweeping slowly across the galaxy's position to enhance its visibility with motion. Once you have the Pinwheel in view, use averted vision. Do not look directly at the galaxy but off to one side instead. This will allow the light to fall on a more sensitive area of your eye's retina.

At 47×, M33 shows a weakly brighter core measuring 30′ by 13′ and running north to south with a slightly brighter, east-to-west inner core. The outer halo is extremely faint, about 52′ by 28′, and runs north-northeast to south-southwest.

The Pinwheel Galaxy contains **NGC 604** — one of the grandest starforming regions known. It is nearly 100 times larger across than our galaxy's vast Orion Nebula (see page 46). This starbirth nest is 12′ northeast of M33's nucleus and 1.1′ northwest of a 10.9-magnitude star. The pair resembles a double star with one component that won't quite come to focus. Through my 4.1-inch scope at 68×, NGC 604 looks like a very small smudge.

Now let's return to our own galaxy, where we'll find the open star cluster **NGC 752** (Caldwell 28). You can sweep it up ⅖ of the way from 3rd-magnitude Beta (β) Trianguli to 4th-magnitude Upsilon (υ) Andromedae. This large cluster is best viewed with an eyepiece that gives

a 1° or larger view. My little refractor at 47× reveals a golden 7th-magnitude star as well as about 90 fainter stars spread loosely across 47′ of sky. Many of the stars seem to be arranged in crisscrossing chains. A pair of 6th-magnitude stars lies outside the group's south-southwest border with one appearing golden and the other orange. With increasing aperture, many additional stars in this cluster reveal subtle shades of yellow, orange, and blue.

Andromeda's Realm

Object	Type	Mag.	Dist. (l-y)	RA	Dec.
M31	Spiral galaxy	3.5	2.5 million	0ʰ 42.7ᵐ	+41° 16′
M32	Elliptical galaxy	8.2	2.5 million	0ʰ 42.7ᵐ	+40° 52′
M110 (NGC 205)	Elliptical galaxy	8.0	2.5 million	0ʰ 40.4ᵐ	+41° 41′
M33	Spiral galaxy	5.7	2.7 million	1ʰ 33.9ᵐ	+30° 39′
NGC 604	H II region	—	2.7 million	1ʰ 34.5ᵐ	+30° 48′
NGC 752	Open cluster	5.7	1,300	1ʰ 57.8ᵐ	+37° 41′

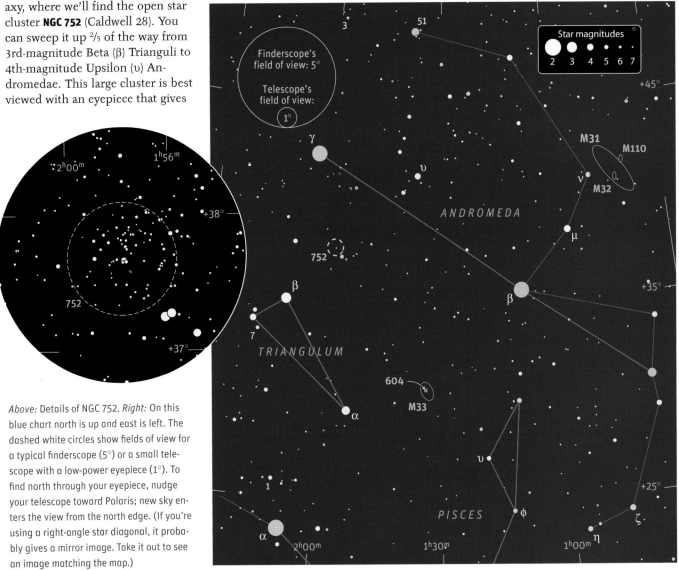

Above: Details of NGC 752. *Right:* On this blue chart north is up and east is left. The dashed white circles show fields of view for a typical finderscope (5°) or a small telescope with a low-power eyepiece (1°). To find north through your eyepiece, nudge your telescope toward Polaris; new sky enters the view from the north edge. (If you're using a right-angle star diagonal, it probably gives a mirror image. Take it out to see an image matching the map.)

Sculpting at the South Galactic Pole

YOU CAN SEE SCULPTOR low in the south on the all-sky chart on page 31. It is a relatively recent constellation, one of 13 introduced by the French astronomer Nicolas-Louis de Lacaille to commemorate the instruments of art and science. Sculptor first appeared on a chart in the *Mémoires* of the Royal Academy of Sciences for 1752 (published in 1756) as L'Atelier du Sculpteur, the Sculptor's Workshop. The chart depicted a carved bust on a platform, a table, and three sculpting tools.

Skygazers will be hard-pressed to see the sculptor's studio among the stars. Sculptor contains only six stars of magnitude 5.0 and brighter, none bearing common names. The constellation's brightest star is 4.3-magnitude Alpha (α) Sculptoris. You can pick it out in the night sky by noting that the brighter stars Iota (ι) and Beta (β) Ceti point toward it.

We'll begin our small-scope tour at blue-white Alpha, first moving 1.7° north-northwest to a 6th-magnitude reddish orange star, then continuing along the same line another 1.4° to the globular cluster **NGC 288.** It's

Left: The mottled glow of the galaxy NGC 7793 has no easy stars around it to serve as guideposts, for none brighter than 11th magnitude appears in this 90-minute exposure at the f/10 focus of an 8-inch Schmidt-Cassegrain telescope. Jeffrey L. Jones of Tracy, California, took it on September 1, 1989. *Right:* Photographs show stars in the globular cluster NGC 288 more readily than a telescope used visually. Eugene A. Harlan took this 50-minute exposure on September 30, 1962, using his 6-inch f/6 astrocamera. The field shown is 0.7° wide, and the star above right of center is magnitude 8.5. All illustrations in this article have north up.

The huge size of the galaxy NGC 253 is apparent from the fact that the 9th-magnitude star off the oval's top edge is ⅓° from the pair of similar stars below center. Jerry Lodriguss made this hour-long exposure on October 10, 1994, with an Astro-Physics 130-millimeter f/8 StarFire refractor.

easy to spot with my 4.1-inch (105-millimeter) refractor at low power as a small, round glow. Boosting the magnification to 87× makes NGC 288 more obvious. The cluster appears a little brighter toward the center, slightly mottled, and about 8' across. Noted observer Walter Scott Houston was able to resolve some of the group's halo stars in giant 5-inch binoculars at 20×.

Just 37' south-southwest of NGC 288 lies the south galactic pole, one of the two spots in the sky farthest from the starry, dust-laden plane of our galaxy's disk. Here we have a clear window into the depths of the universe, where we can view distant galaxies unobscured by the Milky Way. One of the most impressive is **NGC 253,** sometimes called the Silver Coin Galaxy and also known as Caldwell 65.

You can find the Silver Coin by scanning 1.8° northwest of NGC 288. It is visible through binoculars or a large finder in a dark sky. This galaxy is fairly bright in my 4.1-inch scope, and at 87× I can trace it out to 20' by 4' running northeast to southwest. It is broadly brighter along its long axis and subtly dappled with light and dark patches. A pair of 9th-magnitude stars hug the southeastern side.

Caroline Herschel discovered NGC 253 in 1783 while sweeping the sky for comets, a noteworthy accomplishment since this galaxy never climbed higher than 12° above the horizon at her observing site in England. NGC 253 is a barred spiral galaxy, highly inclined to our line of sight. It is the brightest member of a small cluster of galaxies known as the Sculptor Group. At a distance of 10 million light-years, this is the closest galaxy cluster to the Local Group (the one in which our galaxy resides). The Silver Coin is also the closest starburst galaxy, a title it earns by having an exceptionally high rate of star formation in its core.

Now let's turn attention to the 4.6-magnitude star marked δ, for **Delta Sculptoris,** near the center of the chart above. Notice that Delta Scl lies about two-fifths of the way from the bright star Fomalhaut to Beta Ceti and a little below the line between them. It is the brightest star in the area and a low-power double for a small telescope. The white primary has a 9th-magnitude secondary, 74″ to the west-northwest, but it's too dim for me to make out any color. This companion's spectral class is generally given as G (yellow), but it is described in the Webb Society's *Visual Atlas of Double Stars* as "intensely blue." Do you see any color here?

After visiting Delta, star-hop 3.3° east-southeast to the 5.0-magnitude star Zeta (ζ) Sculptoris. Zeta lies in the foreground of the large open cluster **Blanco 1.** With my 4.1-inch scope at 17× I see a dozen fairly bright stars, mostly arranged in a **V** shape with one curved arm. Another dozen faint stars are scattered in and around the degree-wide **V,** swelling its apparent size to about 1.4°.

Although Blanco 1 does not stand out well from surrounding stars, it is a true cluster. The group was discovered in 1949 by Victor M. Blanco, who noticed that the area contains five times the usual concentration (at similar galactic latitudes) of stars brighter than 9th magnitude of spectral type A0. Recent studies have shown that this is a young star cluster perhaps 90 million years old, similar in age to the Pleiades. It lies about 880 light-years away and may have as many as 200 members. Its unusual position high above the galactic plane suggests

For deep-sky viewing this month, check out the constellation of Sculptor just east of the 1st-magnitude star Fomalhaut. Appearing rather vacant to the naked eye, it has much to offer the owners of small telescopes.

that Blanco 1 may have formed by a different mechanism than other nearby clusters.

From Zeta Scl in Blanco 1, we'll move exactly 3° south-southwest to our final deep-sky treasure, **NGC 7793.** This galaxy is never more than 14° above my horizon. Even though it's embedded in the skyglow from a nearby city, I find NGC 7793 to be reasonably obvious through my 4.1-inch scope. At 87× it looks fairly faint overall with a slightly brighter, small center. The 6′-by-4′ oval is aligned nearly east-west.

NGC 7793 is another spiral galaxy of the Sculptor Group. It displays short, asymmetric arms instead of the long, graceful arms seen in classic spirals. Such arms are probably created by ephemeral star-forming regions that are stretched into spiral-like shreds by galactic rotation. These shreds are short-lived in astronomical terms — about 100 million years.

Sculptor contains several other galaxies suitable for small-scope viewing. Given my viewing location in Upstate New York, I have picked out only the brightest of those in the northern reaches of the constellation. More southerly observers might want to take out a good atlas and look for some of the other treasures held by the constellation at our galaxy's south pole.

Highlights of Sculptor

Object	Type	Mag.	Size/Sep.	Dist. (l-y)	RA	Dec.
NGC 288	Globular cluster	8.1	12′	27,000	0ʰ 52.8ᵐ	−26° 35′
NGC 253	Galaxy	7.6	26′ × 6′	10 million	0ʰ 47.6ᵐ	−25° 17′
δ Scl	Double star	4.6, 9.3	74″	144	23ʰ 48.9ᵐ	−28° 08′
Blanco 1	Open cluster	4.5	89′	880	0ʰ 04.2ᵐ	−29° 56′
NGC 7793	Galaxy	9.3	9′ × 7′	9 million	23ʰ 57.8ᵐ	−32° 35′

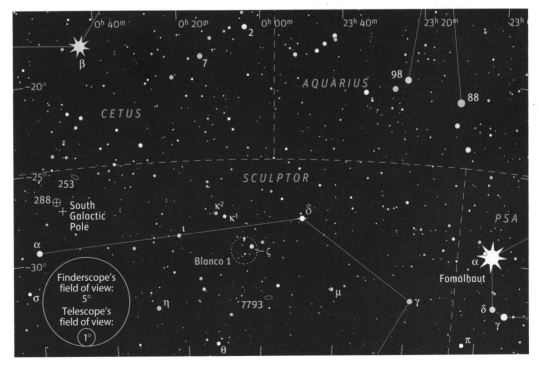

ATLASES

Cambridge Star Atlas, 3rd Edition
Wil Tirion (Cambridge University Press)

Here's a basic atlas that's a good choice for skygazers who are still learning their way around the constellations. It contains a Moon map, monthly sky maps, and shows stars to magnitude 6.5 (plus nearly 900 deep-sky objects). A table of interesting telescopic targets accompanies each chart in the atlas.

MegaStar 5
(Willmann-Bell)

This software atlas has more than 208,000 deep-sky objects, many of which can be overlaid with photographic images. Stars can be plotted from any of three included star atlases, which contain more than 15 million stars. The easy-to-use filters help you tailor the charts to your needs. *(Requires Windows 95 or higher, 32 megabytes RAM, 40 megabytes hard-disk space, CD-ROM drive.)*

Norton's Star Atlas and Reference Handbook, 20th Edition
Ian Ridpath, editor (Pi Press)

It's similar in scope to the *Cambridge Star Atlas,* though it lacks the monthly sky maps. It does include a lengthy astronomy handbook.

Sky Atlas 2000.0, 2nd Edition
Wil Tirion and Roger W. Sinnott (Sky Publishing Corp.)

This is a good atlas to use in conjunction with this book. It plots stars to magnitude 8.5 as well as 2,700 deep-sky objects. Included are handy close-up charts of some of the crowded regions of the sky.

Sky Atlas 2000.0 Companion, 2nd Edition
Robert A. Strong and Roger W. Sinnott
(Sky Publishing Corp.)

The text lists and briefly describes every object plotted on *Sky Atlas 2000.0.*

Sky & Telescope's Pocket Sky Atlas
(Sky Publishing Corp.)

Here's a new star atlas that's great to use at the telescope because it's compact (6 by 9 inches) and spiral bound. It has 80 charts, more than 30,000 stars to magnitude 7.6, and some 1,500 deep-sky objects. Also included are close-up charts of the Orion Nebula region, Pleiades, Virgo Galaxy Cluster, and Large Magellanic Cloud.

Sky & Telescope's Star Wheel
(Sky Publishing Corp.)

This planisphere is a simplified "atlas" that will show you which constellations are visible on any date and time. It's available for different latitudes.

Uranometria 2000.0 Deep Sky Atlas
Wil Tirion, Barry Rappaport, and Will Remaklus
(Willmann-Bell)

This is an advanced star atlas with stars down to magnitude 9.75 and more than 30,000 non-stellar objects. It's great for pinning down those faint objects. It includes both 5th- and 6th-magnitude star charts to help you navigate the main atlas, has 26 close-up charts of crowded regions of the sky, and the index gives the chart numbers for all Messier, NGC, and IC objects (as well as those for bright stars and objects with common names). *Uranometria 2000.0* comes in two overlapping volumes — the northern sky and the southern sky.

Uranometria 2000.0 Deep Sky Field Guide
Murray Craigin and Emil Bonanno (Willmann-Bell)

This text contains useful data and a chart index for all the deep-sky objects included in *Uranometria 2000.0.*

OBSERVING GUIDES

Deep-Sky Companions: The Messier Objects
Stephen James O'Meara (Sky Publishing Corp.)

Eke all the detail you can out of the Messier objects with the help of this detailed guide.

Deep-Sky Companions: The Caldwell Objects
Stephen James O'Meara (Sky Publishing Corp.)

Reach beyond the Messier objects with this in-depth look at some of the sky's most interesting deep-sky delights.

Deep-Sky Wonders
Walter Scott Houston, edited by Stephen James O'Meara
(Sky Publishing Corp.)

Here are selections from 48 years of the *Sky & Telescope* column of the same name penned by Walter Scott Houston, who charmed us all with his journeys into the night sky.

Observing Handbook and Catalogue of Deep-Sky Objects
Christian B. Luginbuhl and Brian A. Skiff
(Cambridge University Press)

This excellent catalog describes more than 2,000 deep-sky objects (in constellations north of declination –50°) as seen through 2.4-inch to 12-inch telescopes.

The Night Sky Observer's Guide
George Robert Kepple and Glen W. Sanner
(Willmann-Bell)

This two-volume guide to more than 5,500 double stars, variable stars, and non-stellar objects in 64 constellations visible from the Northern Hemisphere includes descriptions and sketches of objects as seen through 2-inch to 22-inch telescopes. However, note that many sights listed are not visible through a small telescope.

Touring the Universe Through Binoculars
Philip S. Harrington (Wiley)

There are more than 1,100 deep-sky objects listed in this guide, and more than 400 are described in some detail. Most of these sights are great for small scopes too!

Turn Left at Orion
Guy Consolmagno and Dan M. Davis
(Cambridge University Press)

A fine observing guide to use with your small telescope. It lists 100 deep-sky sights and includes sketches of how they appear in the eyepiece.

GENERAL REFERENCE
NightWatch: A Practical Guide to Viewing the Universe, 3rd Edition
Terence Dickinson (Firefly)

This is one of the best introductions to observing the night sky ever written, delightfully touching on a wide range of astronomical pursuits.

Starlight Nights
Leslie C. Peltier (Sky Publishing Corp.)

This charming and inspirational book chronicles Peltier's astronomical adventures that began when he glimpsed the Pleiades through the window of his family's Ohio farmhouse in 1905.

The Backyard Astronomer's Guide, Revised Edition
Terence Dickinson and Alan Dyer (Firefly)

Here's an in-depth look at the world of amateur astronomy, including the ins and outs of equipment.

The Backyard Stargazer
Pat Price (Quarry Books)

This friendly, conversational beginner's guide to exploring the night sky includes 46 fun and practical observing projects to get you started.

WEB SITES OF INTEREST
American Association of Variable Star Observers (AAVSO)
www.aavso.org/aavso

This site contains a wealth of information on variable stars including comparison charts and light curves.

Cartes du Ciel (Sky Charts)
www.stargazing.net/astropc

This is a software atlas that can be downloaded for free.

Internet Amateur Astronomers Catalog (IAAC)
www.visualdeepsky.org

Read observations by other amateur astronomers with a wide range of observing equipment and different levels of experience; post your own observations here too!

Messier45.com
http://messier45.com

The site contains information for about 500,000 deep-sky objects and more than 2 million stars. It includes maps and images for every object and has powerful search capabilities.

SEDS — Students for the Exploration and Development of Space
www.seds.org

SEDS hosts many fascinating astronomical Web sites (including Messier and NGC catalogs) that contain amazing amounts of information.

Sky Publishing Corporation
SkyandTelescope.com
NightSkyMag.com

These two sites support *Sky & Telescope* and *Night Sky*, the new bimonthly magazine for beginners, and contain observing articles and how-to tips.

The Washington Double Star Catalog
http://ad.usno.navy.mil/wds/wds.html

This catalog contains data for over 100,000 star systems and is the world's principal database of double- and multiple-star information.

Every deep-sky object described in the text is included here as it is listed in the table that accompanies each essay. Many sights (for example M8 and M63) also have popular names listed as crossreferences (Lagoon Nebula, Sunflower Galaxy). There is no list of generic entries for each type of deep-sky object. Page numbers in **boldface** refer to photographs or illustrations.

Star or object designations such as Epsilon Bootis, 61 Cygni, or R Scuti, which contain the genitive form of a constellation name, are indexed under the constellation, in these cases Boötes, Cygnus, and Scutum, respectively. Those double stars designated with the Greek letter sigma (Σ) are listed at the end of the index.

INTRODUCTION

p10 Akira Fujii; **p11** Richard Tresch Fienberg; **p12** [top] Steve Cannistra, [bottom] *Sky & Telescope;* **p13** *Sky & Telescope* (×2); **p14** Dennis di Cicco; **p15** [top] Alan Adler, [bottom] Dennis di Cicco(×2); **p16** [top left] Alan Dyer, [top right] Craig Michael Utter, [bottom] Becky Ramotowski; **p17** David Kone.

HOW TO USE THE ALL-SKY STAR MAPS

p18 *Sky & Telescope* (×2); **p21–30** *Sky & Telescope.*

All finder charts in Chapters 1 through 4 inclusive: *Sky & Telescope.*

CHAPTER 1

p34 [top] Akira Fujii, [bottom] Martin C. Germano; **p36** [top] William C. McLaughlin, [bottom] Jon A. Kolb; **p38** [top] Kim Zussman, [bottom] Robert Gendler; **p40** [top] Bobby Middleton, [bottom] Robert Gendler; **p42** [top] Akira Fujii, [bottom] Bob & Janice Fera; **p44** Russell Croman; **p46** Steve Cannistra; **p47** Lick Observatory; **p48** [top] George R. Viscome, [bottom] Akira Fujii; **p50** Akira Fujii; **p52** Lee C. Coombs; **p53** Preston Scott Justis (×2); **p54** [top] Till Credner/ Sven Kohle, [bottom] Akira Fujii; **p55** Sean Walker; **p56** [top] Martin C. Germano, [bottom] Chris Cook; **p58** Akira Fujii, [inset] R.B. Milton; **p59** Martin C. Germano (×3); **p60** George R. Viscome (×2); **p62** Akira Fujii; **p63** Akira Fujii.

CHAPTER 2

p66 Robert Gendler; **p67** Akira Fujii; **p68** [top right] Bill & Sue Galloway, Adam Block, NOAO/AURA/NSF, [top left] Dan Stotz, Mike Ford, Adam Block, NOAO/ AURA/NSF, [bottom] George R. Viscome; **p70** [top] STScI/Digitized Sky Survey, [bottom] Bobby Middleton; **p72** [top] Akira Fujii, [bottom] Richard D. Jacobs; **p74** [top] Akira Fujii, [bottom left] Akira Fujii, [bottom right] Gérard Therin; **p75** Sue French; **p76** Akira Fujii (×2), [b&w insets] Sue French (×2); **p78** [top] George R. Viscome, [bottom] Martin C. Germano; **p79** Martin C. Germano; **p80** Bob & Janice Fera (×2); **p81** Sue French; **p82** [upper left] Martin C. Germano, [upper right] John Mirtle, [lower left] John Mirtle, [lower right] Martin C. Germano; **p84** Adam Block, NOAO/ AURA/NSF; **p86** Sissy Haas (×3), [bottom] Akira Fujii;

p87 Joseph Liu; **p88** Akira Fujii; **p90** [top] Preston Scott Justis, [bottom] Akira Fujii; **p91** Akira Fujii, [inset] Dennis di Cicco; **p92** *Sky & Telescope* (×6); **p93** *Sky & Telescope* (×4); **p94** George R. Viscome.

CHAPTER 3

p98 [top] Bob & Janice Fera, [bottom] Preston Scott Justis; **p99** Martin C. Germano; **p100** Akira Fujii; **p102** Adam Block, NOAO/AURA/NSF (×2); **p104** [top] Palomar Observatory Sky Survey, [top, inset] Tan Wei Leong, [bottom] Palomar Observatory Sky Survey; **p106** [top] AAVSO, [bottom left] Evered Kreimer, [bottom middle] Evered Kreimer, [bottom right] Martin C. Germano; **p108** [top] Digitized Sky Survey, [bottom] Akira Fujii; **p110** [top] Sue French, [bottom] Akira Fujii; **p111** Sue French; **p112** [top] Sue French, [bottom] Luke Dodd; **p113** Jose Torres; **p114** [top] Martin C. Germano, [bottom] Lee C. Coombs; **p116** Akira Fujii; **p118** [upper left] Martin C. Germano, [upper right] Lee C. Coombs, [middle left] Digitized Sky Survey, courtesy U.K. Schmidt Telescope Unit and NASA/ AURA/STScI, [middle right] Evered Kreimer, [lower left and right] Evered Kreimer; **p120** Akira Fujii, [inset] Alan MacRobert; **p122** Akira Fujii; **p123** Akira Fujii; **p124** Akira Fujii (×2); **p126** [top left] Ronald P. Manley, [top right] Sue French, [bottom] Michael Stecker.

CHAPTER 4

p130 [top] Sue French, [bottom] Ben Gendre; **p132** [top] Sue French, [bottom] Bobby Middleton; **p134** [left] Akira Fujii, [right] Martin C. Germano; **p136** [top left] Adriano Defreitas, [top right] Bob & Janice Fera, [bottom] Akira Fujii; **p138** Akira Fujii (×2); **p140** [top] Martin C. Germano, [bottom] Akira Fujii; **p141** Martin C. Germano; **p142** [top] Chip Levinson, Adam Block, NOAO/AURA/NSF, [bottom] William C. McLaughlin; **p144** [top] Bill Ferris, [bottom] Gary L. Stevens; **p146** Akira Fujii, [b&w inset] Sue French; **p147** Akira Fujii (×2); **p148** Jerry Lodriguss; **p150** Akira Fujii; **p151** George R. Viscome (×3); **p152** [top] George R. Viscome, [bottom] Sean Walker; **p153** George R. Viscome; **p154** [top] Preston Scott Justis, [bottom] Akira Fujii; **p155** Preston Scott Justis; **p156** Robert Gendler; **p158** [top left] Jeffrey L. Jones, [top right] Eugene A. Harlan, [bottom] Jerry Lodriguss.